The Urban Book Series

Aims and Scope

The Urban Book Series is a resource for urban studies and geography research worldwide. It provides a unique and innovative resource for the latest developments in the field, nurturing a comprehensive and encompassing publication venue for urban studies, urban geography, planning and regional development.

The series publishes peer-reviewed volumes related to urbanization, sustainability,urban environments, sustainable urbanism, governance, globalization, urban and sustainable development, spatial and area studies, urban management, urban infrastructure, urban dynamics, green cities and urban landscapes. It also invites research which documents urbanization processes and urban dynamics on a national, regional and local level, welcoming case studies, as well as comparative and applied research.

The series will appeal to urbanists, geographers, planners, engineers, architects, policy makers, and to all of those interested in a wide-ranging overview of contemporary urban studies and innovations in the field. It accepts monographs, edited volumes and textbooks.

More information about this series at http://www.springer.com/series/14773

Nasrin Banu · Shahab Fazal

Livelihood and Wellbeing in the Urban Fringe

Springer

Nasrin Banu
Department of Geography
Aligarh Muslim University
Aligarh
India

Shahab Fazal
Department of Geography
Aligarh Muslim University
Aligarh
India

ISSN 2365-757X ISSN 2365-7588 (electronic)
The Urban Book Series
ISBN 978-3-319-39659-0 ISBN 978-3-319-39660-6 (eBook)
DOI 10.1007/978-3-319-39660-6

Library of Congress Control Number: 2016940316

Printed on acid-free paper

This Springer imprint is published by Springer Nature
The registered company is Springer International Publishing AG Switzerland

Contents

List of Figures

List of Tables

Introduction

Overview

Our planet is becoming a planet of cities as the share of global urban population has increased from 15 % to more than 50 % in the last century (Fazal 2013). Urban population around the world has grown 1.6 times more than rural population since 1950, driven both by migration from the countryside and by natural growth. The trend shows that urbanization process has now spread predominantly among rural developing countries (UNFPA 1996). The cities of developing countries will account for most of the urban growth while urban population in developed would remain more or less stagnant (Figs. 1, 2 and 3).

Rapid growth of urban population in Asian countries during a few decades in the second half of the last century has led to the speculation of urban explosion in the region. At the same time, the urbanization level has almost stabilized in developed countries. Researchers have indicated that the cities of developing countries will account for nearly 95 % of this growth. Asian countries will account for 55 % of the world's urban population.

Among the populous Asian countries, India and China alone will account for more than 64 % of the overall growth of urban populations in Asia and a 42 % share of global urban population growth from 2005 to 2025 (Sankhe et al. 2011) (Fig. 4). China, the most populous country of the world, will also have the world's largest urban population at triple the size of that of the United States; India, the second most populous country, also have an urban population double that of the United States (ibid.). In 2011, there were three 10 million plus cities in India, while in 1950 while there was none in 1950. This fact ensures an 'unprecedented urban growth' in India. Over the years there has been continuous concentration of population in class I towns. India shares most characteristic features of urbanization with the developing countries due to its large demographic weight and dynamics of urbanization during the last three decades, where the population increased from 161.6 million in 1981 to 388.3 million in 2011.

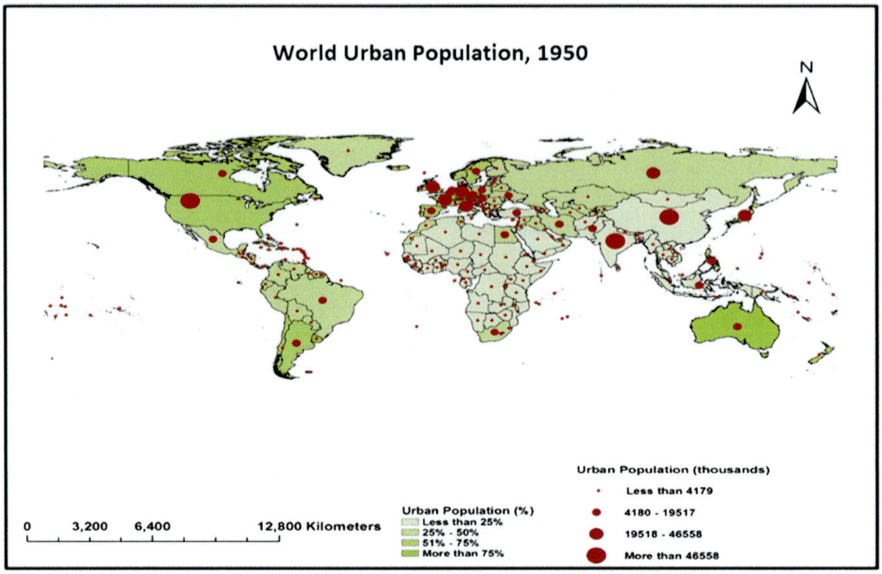

Fig. 1 Distribution of World Urban Population, 1950. *Source* Based on United Nations, Department of Economics and Social Affairs, Population Division

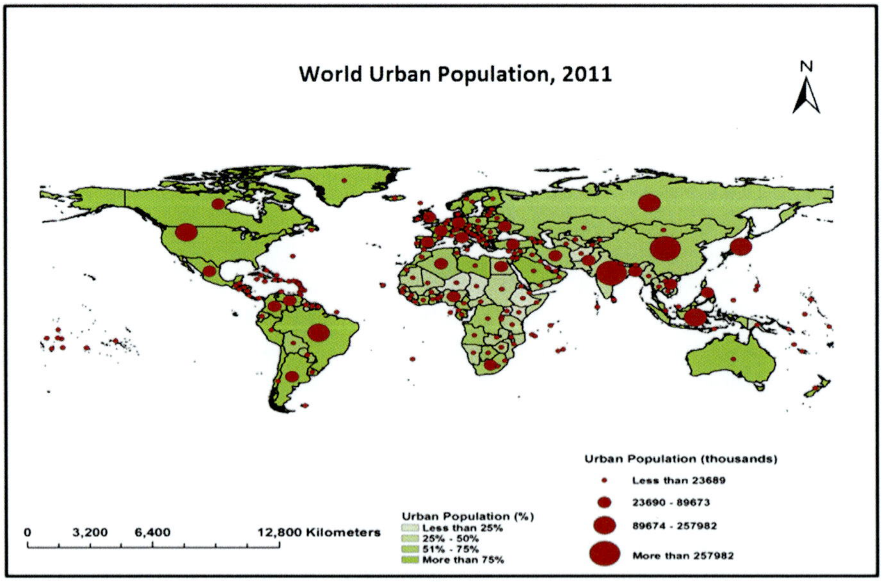

Fig. 2 Distribution of World Urban Population, 2011. *Source* Based on United Nations, Department of Economics and Social Affairs, Population Division

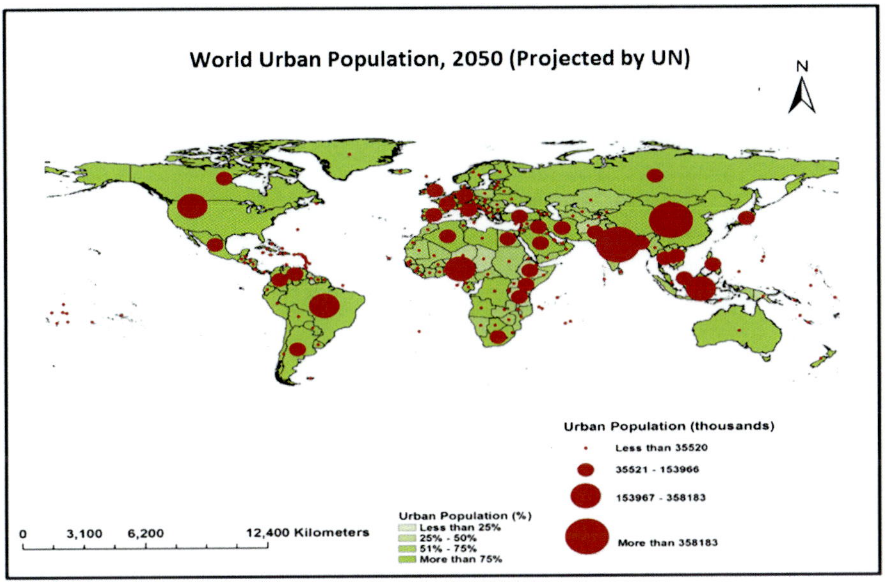

Fig. 3 Distribution of Projected World Urban Population, 2050. *Source* Based on United Nations, Department of Economics and Social Affairs, Population Division

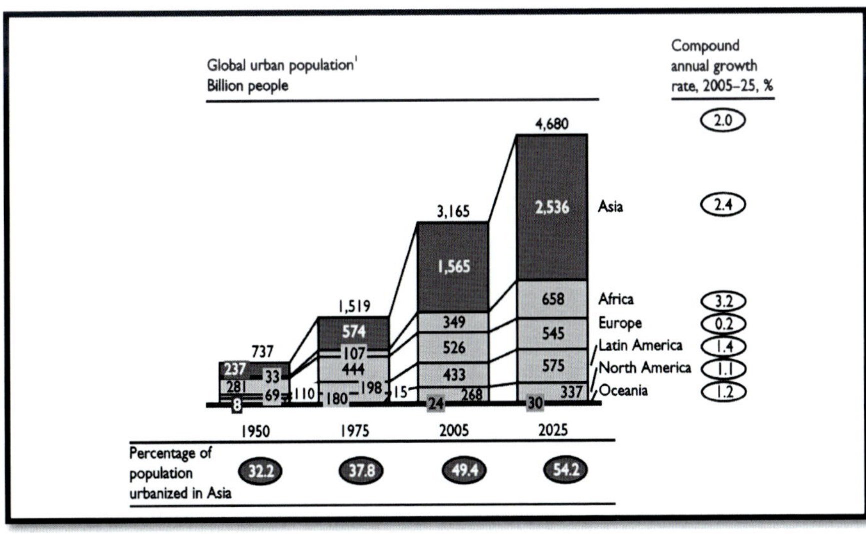

Fig. 4 Global urban population growth. *Source* United Nations; McKinsey Global Institute analysis

The process of urbanization has traditionally been understood as a natural byproduct of economic development. Urbanization takes a variety of forms. It used to be confined chiefly to a gradual expansion of cities. Improved transportation has altered the pattern to include rapid urbanization of land along arterial highways, encirclement of agricultural land, and an uneven diffusion of the urban uses over the countryside. This urbanization process in India is marked by increases in population and expanding of urban bodies by engulfing into their territory (Singh 2002). The spilling over of population from the city into the peripheral areas has been occurring quite rapidly since 1980s in almost all large cities in India (Shaw 2005). This process has resulted also in the emergence of Urban Fringe (UF) which is conceptualized as zone of transition in land use, social and demographic characteristics between continuously built-up urban areas and rural hinterland.

Background of the Study

The present study is undertaken on a relatively smaller city as they have not been given sufficient attention—both in terms of their needs for services and facilities, such as piped water supplies, provision for services and facilities and for measures to strengthen the basis for their livelihood. Urban fringe is considered as a transitional zone between built-up urban areas and rural hinterland. Urban fringe is found to be neither under the control of urban authorities nor under the control of rural authorities (Adu-Ampong et al. 2008).

Although Aligarh is a relatively small city, compared to the other cities for which literature is available (most of them are for metropolitan cities) it is also a fast growing city in India. The issues examined here are certainly not limited to Aligarh city and many of the features can be traced for other cities as well. Like many other cities in north India, Aligarh is expanding fast, moreover, it is surrounded by a populous rural area with productive and rich agricultural hinterland. Such conditions give rise to many conflicts between rural and urban economies, values and people. However, they can also generate mutually beneficial complementarities in the rural and urban spheres. The net outcome is highly selective for different groups and different locations.

The study of livelihood transformation in urban fringe is relevant for understanding the changes occurred in the urban fringe of Aligarh city over last two to three decades. It is important to understand the adaption by households in the changing economic-, social- and cultural environment of the urban fringe areas due to urbanization processes. The livelihood approach is not just a response to modernization perspectives, but it is also for analyzing the developmental process operating in the urban fringe since 1980s.

The expansion of the city strongly influences the villages in terms of land use and population, both physically and in a socio-economic sense. Its influence stretches far beyond the immediately adjacent area. This study concentrates on the livelihood of the villagers who have lived there for generations as well on that of the migrants residing in the villages. Livelihood according to the *Oxford English Dictionary* refers to a 'manner of life' but also to 'the various means of maintenance'; it suggests that it is a method to acquire food, income, property or inherit resources. Thus this study considers livelihood as all the activities forming the means of generating a living; the notion of livelihood is narrowed down here to reflect social and economic household variables (during the course of sample survey one respondent used the term *Grahasthi*); of these, especially the occupational characteristics are studied thoroughly.

A fundamental characteristic of urban fringe livelihood in contemporary developing economies like India is their ability to adapt in order to survive. Thus the definition of livelihood also includes change over time and adaptation to evolving circumstances. The livelihood therefore is an ongoing process in which the elements do not remain static but varies from one season or from one year to the next. The assets can be built up, eroded or instantaneously destroyed as available activities and access to resource and opportunities may fluctuate seasonally and across the year.

With modernization and transformation of economic activities, the traditionally dominant agricultural activities are under increasing pressure, although new possibilities are emerging for market-oriented agriculture. There are great differences in access to resources and opportunities among villagers, leading to interesting new patterns of livelihood. For non-agricultural occupations, the villagers' options are enhanced by their proximity to the expanding city. Land use and occupational trends are highly interrelated. For example, Brookfield et al. (1991) found that in Kuala Lumpur's fringe there was migration of working population and the land acquires an aspect of neglect. Other sources emphasize the stimulating effects on agriculture (e.g. Mulay and Ray 1973; Hill 1986). This inspired us to investigate to what extent the transformation of agriculture in urban fringe has taken place. A number of other consequences of urbanization—namely the pace and direction of city expansion, land markets and environmental concerns—are also treated in this study.

In order to study the socio-economic impact, this work has used a more concrete operationalization in the view of the manifestation of urbanization at micro level. Some indications of its scope were derived from the various descriptions in the literature (e.g. Ramachandran 1989; McGee 1991; Nangia 1976 & Thakur 1991). But exploratory fieldwork was needed to find leads to relevant aspects of urbanization at the village level in the study area. For the villages near Aligarh city, it was necessary to translate terms with a profoundly 'Western' bias into their Hindi-equivalent. *Nagarikaran* (literally means urban development) proved to be the most appropriate term to explain the purpose of the study during the interviews, as it evoked references to the changes in the lives of households and their villages as a result of urbanization.

The following aspects are supposed to be directly or indirectly caused by urbanization: land use changes in favour of urban types of land use, value addition to the products, diversification in their economic activities, location of urban-related activities in or near the village; better access from the village to the urban areas and vice versa; and expansion of urban types of amenities and infrastructure. All these processes have far-reaching consequences for the livelihood of the urban fringe population, as they lead to a diversification of sources of income among the villagers, to different uses of the agricultural land, and to change in living conditions. A period of about 30 years, approximately one generation, was chosen as the time frame for two reasons.

- First, there are some relevant secondary sources that provide insight into local history and the situation in the 1980s.
- Second, this time span corresponds with the period of reliable 'recall' by most respondents and is manageable for the researcher.

The selected area encompasses a wide range of villages: from places that have remained relatively rural, deriving an income mainly from agriculture, to villages that have been entirely enveloped by urbanization. The distance from city, land use pattern and the transformation of the occupational structure are the most important criteria for clustering and selecting villages for further study. A basic household survey was conducted to describe the situation in 44 villages (Fig. 5).

Objectives

The present work on immediate urban centre has two main objectives. The first is to provide a better understanding of how urban force acts to shape or mould the urban fringe of Aligarh city over time, and their impact on economic activities and social development.

- General objectives
  To evaluate the features of livelihoods and their transitions in the urban fringe of Aligarh city.
- Specific objectives
  - To identify the changes in livelihood activities and modified livelihood option.
  - To find the impact of city proximity on

    Social environment in the urban fringe
    Economic environment in the urban fringe
    Status of infrastructural facilities in the urban fringe
    Physical environment in the urban fringe

Methodology and Data Sources

In order to draw upon a comprehensive range of information, the study makes use of a variety of resources and methods. The study is mainly based on primary data derived from the field survey[1]. The work also uses secondary data.

Secondary Data

These include statistical information of the city and the various settlement forms and their expansion in the regions and field areas concerned. Census data provided useful information at village level and more generally, about the land, population and economic activities of people in the area. Census data were used to categorize the sample villages within the urban fringe on the basis of land use and occupational characteristics. The village level secondary data were collected through village revenue officer (*Patwari*) who keeps records on village land, including its use and ownership. The aggregated primary data on crop patterns at village level was used to supplement the village description. The aggregated data, collected from Agricultural Census on cropping pattern, leasing land, input, sources of irrigation, etc., at tehsil level is used to support the primary data. From the village directory, the data on amenities and facilities has been collected here over last three decades to evaluate the changes over the period of time and to show the infrastructural development at village level the data was collected from Economics and Statistics Division, Sankhyikiya Patrika (Uttar Pradesh). Disaggregated data at household level were used to analyze land ownership and transactions. Some additional background information about the villages was derived from various statistical bulletins and governmental publications, as they provided a historical perspective.

The study, which basically integrated the primary and secondary data, assess the primary influences of urbanization processes on land use, demographic, occupation and socio-economic condition of fringe villagers of Aligarh city. Data collected from household level, defined as "those living and eating together", excludes family members who live elsewhere, which means household is a unit in which family members share their food and shelter (Preston 1994). Elder male member of the family has been considered as head of the family and most of the time head of the family member was chosen as respondents for questioning.

[1]The survey was conducted in 2013.

Primary Data

Primary information regarding social, economic and livelihood structure was collected from respondents using a well-structured questionnaire. After preparing the questionnaire the pilot survey of 2–3 villages was done to get the real idea of the study area and for the modification in the questionnaire. The primary data was collected from stratified random sampling method; where on the basis of distance three buffer zones at 5 kms distance were drawn from the city municipal boundary. They were named as immediate urban fringe zone (villages located in 0–5 kms from city municipal boundary), intermediate urban fringe zone (villages located between 5–10 kms from city municipal boundary) and distant urban fringe zone (villages located in 10 kms and beyond the city municipal boundary). The selection was further made considering the following indicators—accessibility, population size, economic activity and existing infrastructure facilities. From the each zone 10 % villages were selected and a total of 44 villages sampled for in-depth analysis, 11 villages from immediate urban fringe zone, 15 villages from intermediate urban fringe zone and 18 villages from distant urban fringe zone. Of these 44 villages about 10 % households were sampled, where again the selection of household was made giving representation to all section of the population. A total of 757 households were sampled, 393 households from villages of immediate urban fringe zone, 198 from villages of intermediate urban fringe zone and 166 from villages of distant urban fringe zone (Table 1 and Table 2).

Table 1 Aligarh urban fringe: sampled villages

Immediate Urban Fringe (Up to 5 km)	**Intermediate Urban Fringe (5–10 km)**	**Distant Urban Fringe (10 km and above)**
1.Maheshpur	1.Barautha	1.Maimari
2.Barhati	2.Jarauth	2.Arrana
3.Wajidpur Nada	3.Kothia	3.Chuharpur
4.Kwarasi	4.Nagaula	4.Ahamadpuri
5.TalaspurKalan	5.BhankriKhas	5.Tejpur
6.RamgarhPanjoopur	6.Digsi	6.SisroiNaglaDhak
7.Dhanipur	7.Imlani	7.Peepali
8.KasbaKol	8.Karsua	8.DaudGarhi
9.Alinagar	9.KishanGarhi	9.Navalpur
10.Sarsaul	10.Jatpura	10.GadaKhera
11.HaivatpurFagoi	11.Alipur	11.NaglaSabalUrfGonda
	12.RafipurSiya	12.Jagatpur
	13.Badwamni	13.Pratappur
	14.Bhartua	14.JangalGarhi
	15.MadanpurChhabila	15.Padki
		16.Daupur
		17.Bhataula
		18.Parauri
11 Villages	**15 Villages**	**18 Villages**

Table 2 Aligarh urban fringe: distribution of sampled villages

Zone	Total Number of Villages	Sampled Villages	Total HH	Sampled HH	Total Population	Sampled Population
Immediate Urban Fringe	93	11	7861	393	48979	2611
Intermediate Urban Fringe	119	15	3960	198	25086	1291
Distant Urban Fringe	156	18	3329	166	21020	1239
Total	**368**	**44**	**15150**	**757**	**95085**	**5141**

The sample survey provides grounds for answering most of the research questions about livelihood transformation in Aligarh city urban fringe. Specifically, it allows us to generalize its effects on—and the response by—the village population. The sample survey not only collected quantifiable variables but also acquired longitudinal elements of the respondents. The focus was on livelihood structure and changes in the households. The questions elicit information on circumstances that influence reactions, choices and strategies. Furthermore, the sample survey covered household histories over the last 30 years. This period more or less coincides with the secondary data analysis. Moreover, it seems to reflect what an average middle-aged respondent can reasonably recall.

Help was also got from local informants including village headmen, property dealers, elderly people who know the village history, local teachers and other well-informed local people. Persons in relevant positions at government departments, planning agencies, academic institutions and local administration have also been consulted. They were particularly helpful when starting the field survey. Later they provided a means to double-check information, and they gave feedback on results. Various people in such positions were interviewed to find out about policy and practice and to appreciate the discrepancy between the two.

Methods

For the data analysis the statistical techniques, simple percentage distribution and cross tabulation are used for the study. Even techniques of correlation are used for the analysis of relationship between two indicators or variables. The spatial analysis was also done mainly using GIS techniques—simple overlying, cross operation, geo-referencing and buffer operations.

Various kinds of techniques are used for processing the data. These are

i. Cross tabulations and cartographic methods
ii. Different kinds of descriptive statistics.
iii. Parametric test, especially one-way between-groups ANOVA with post hoc test and non-parametric, especially chi-square test.
iv. Indices by normalization and proportionate possession weighting technique.

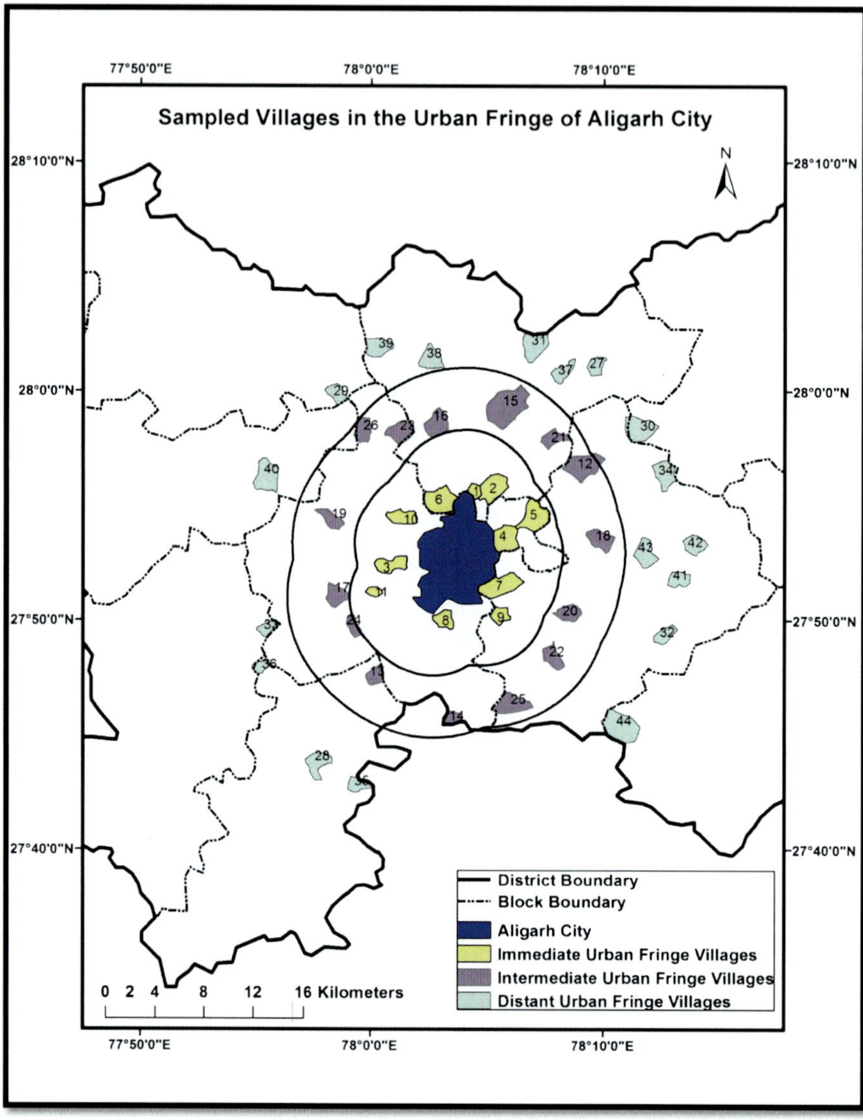

Fig. 5 Aligarh urban fringe: location map of the sampled villages 1. Maheshpur, 2. Barhati, 3. Wajidpur Nada, 4. Kwarsi, 5. Talaspur Kalan, 6. Ramgarh Panjoopur, 7. Dhanipur, 8. Kasba Kol, 9. Alinagar, 10. Sarsaul, 11. Haivatpur Fagoi, 12. Barautha, 13. Jarauth, 14. Kothia, 15. Nagaula, 16. Bhankri Khas, 17. Digsi, 18. Imlani, 19. Karsua, 20. Kishan Garhi, 21. Jatpura, 22. Alipur, 23. Rafipur Siya, 24. Badwamni, 25. Bhartua, 26. Madanpur Chhabila, 27. Maimari, 28. Arrana, 29. Chuharpur, 30. Ahmadpuri, 31. Tejpur, 32. Sisroi NaglaDhak, 33. Peepali, 34. Daud Garhi, 35. Nvalpur, 36. Gada Khera, 37. Nagla SabalUrfGonda, 38. Jagatpur, 39. Pratappur, 40. Jangal Garhi, 41. Padki, 42. Daupur, 43. Bhataula, 44. Parauri.

Apart from those, all sorts of statistical analysis are done by SPSS-17 (Statistical Package for the Social Sciences) software and maps or diagrams are prepared by both Arc GIS- 9.3 and ERDAS-9.1 software.

References

Adu Ampong EF et al (2008) Socio-economic transitions; changing livelihoods in the peri-urban interface, A case study in peri-urban interface of Kumasi. Draft report (July 25, 2008), Tropenbos International, Ghana, Amsterdam University, College of Agriculture and Natural Resource, KNUST

Brookfield H, Hadi AS, Mahmud Z (1991) The city in the village, the in-situ urbanization of villages, villagers and their land around Kuala Lumpur, Malaysia. Oxford University Press, Oxford

Fazal S (2013) Land use dynamics in a developing economy. Regional perspective from India. Springer Publication, The Netherlands

Hill R (1986) Land use change on the urban fringe. Nat Resour 22:24–33

McGee T (1991) The Emergence of desakota regions in Asia: expanding a hypothesis. In: Ginsburg N, Koppel B, McGee T (eds) The extended metropolis: settlement transition in Asia, University of Hawaii Press, Honolulu

Mulay, Ray (1973) Towards modernization: a study of peasantry in rural Delhi. National Pub. House, Overseas Distributors: Books from India, London (Delhi)

Nangia S (1976) Delhi metropolitan region—a study in settlement geography. K.B. Publishers, Delhi

Preston D (1994) Rapid household appraisal: a method for facilitating the analysis of household livelihood strategies. Appl Geogr 14:203–213

Ramachandran R (1989) Urbanisation and urban systems in India. Oxford University Press, Delhi (Reprint 2001)

Sankhe S et al (2011) Urban giants India and China, and their urbanization paths. Environment and urbanization ASIA 2:1–12

Shaw A (2005) Peri-urban interface of Indian cities: growth, governance and local initiatives. Economic and political weekly 40:129–136

Singh Krishna B (2002) Land use systems and changes in India in human dimensions of sustainable development. In: Singh RB (ed) Human dimensions of sustainable development, Rawat Publications, Jaipur

Thakur B (1991) Land utilization and urbanization. The rural-urban fringe. Deep & Deep Publishing House, New Delhi

UNFPA (1996) The State of world population 1996. In: Marshall A (ed) Changing places: population. Development and the urban future, United Nations Population Fund, New York

Chapter 1
Urban Fringe: The Concept

1.1 Concept of Urban Fringe

The term 'urban fringe' appeared for the first time in 1937 when T.L. Smith used it to signify 'the built-up area just outside the corporate limits of the city' (Pryor 1968). This term was widely adopted in the academic literature for the transition zone between city and countryside (Jhonson 1974). The areas surrounding urban centers have many manifestations in the literature in terms of the way it is conceptualized and delimited. Weherwein (1942), in his work on cities in the US used the term rural–urban fringe. It was followed by many other scholars in other countries, including Nangia (1976), Yadav (1987), Ramachandran (1989), Thakur (1991), Lucas and Van Oort (1993) and Fazal (2000). Many other terms are also found to be used in literature for the adjoining areas of urban centers, implying different delimitations and levels of analysis. There has been diversity in the terminology used: urban fringe (Smith 1937; Kabra 1980; Hill 1986 and Kumar 2001), rural-fringe of the city (Leeming and Soussan 1979), the city's countryside (Bryant et al. 1982), Peri-urban fringe (Swindell 1988), desokota regions (McGee 1991), rural hinterland of the city (Kundu 2003), metropolitan fringe (Saini 1989 and Rao 1991, Browder et al. 1995), peri-urban areas (McGee 1991and Dupont 1997), rurban fringe (Schenk 1997) and edge lands (Gant et al. 2011). In this research, the term urban fringe has been used which begins at the point where agricultural land uses appear near the city and extends up to the point where villages have distinct urban land uses and where some persons, at least, from the village community commute to the city daily for work or for other purposes.

"The occurrence of rural—urban fringe is a rather recent phenomenon around Indian cities, though its occurrence around western cities was observed long ago" (Ramachandran 1989). It has been asserted that the rural–urban fringe emerged around the urban centers in India after post-independence (Ramachandran 1989; Siddhartha and Mukherjee 1997). However, there was a radical transformation of urban scene in India. Since 1980s the one lakh cities and million cities began to

N. Banu and S. Fazal, *Livelihood and Wellbeing in the Urban Fringe*,
The Urban Book Series, DOI 10.1007/978-3-319-39660-6_1

grow rapidly. This rapid growth rate of cities could not be accommodate within the cities (ibid.) and spilled it over the adjoining rural areas. The physical expansion of the city brought in itself many changes in economic, social, behavioral aspects of life in the fringe villages, such as change in landuse, cropping pattern and emergence of semiurban society (Siddhartha and Mukherjee 1997).

1.2 Traditional Concepts of the Urban Fringe

The literature on the urban fringe is included in different groups of works dealing with topics such as urban and rural relationships and their links, the extension of metropolitan growth on its surroundings, or changes and transformations in the economic, social, cultural and environmental realms etc. The urban fringe is conceptualized as transitory zone between rural and urban settlements.

One of the literature reviews on the urban fringe, Nottingham and Liverpool Universities (1998) recognize "the complexities of building a spatial framework around what is essentially an amorphous and mobile site for the interaction of various social, economic and cultural processes and inter-linkages between the rural and the urban". Rakodi stressed the relationship between urban and the immediate rural areas being the result of a process over time, the urban fringe is a dynamic zone both spatially and structurally (Rakodi 1998). Here the idea of "shifts" or evolution of the edge of the cities moving "outwards" is clear. The urban fringe seems thus to be considered as the result of urban driven processes more than of territorial processes where rural and urban forces interact. Browder et al. (1995), has underlined it by stating "traditional conceptions of the metropolitan fringe", the term that are preferred to urban fringe, considering it as somewhat specific to the French research literature.

The literature on less urbanized African cities is seen as constituting "the most robust and thematically unified conception of fringe development" (Browder et al. 1995). This trend of research saw the poor or "informal" fringes of the city as spaces that are the product of the interaction of state intervention and policies and the action and practices of the inhabitants seen as everyday use and appropriation of spaces, land and housing strategies and self-building practices.

Similarly, both Latin American and Asian research have also contributed significantly to the development of the conceptual framework for understanding the expansion of the city into the peripheral rural surroundings and the key themes are as follows:

i. The importance of the informal economy in the urban fringe areas, reflected in the proliferation of petty commodity production, multiple job-holding, self-help housing, informal lending, unlicensed activities, etc.
ii. The importance of urban fringe agriculture and rural linkages, such as food brought from the home village of a urban fringe resident, cash income remittance to rural villages, consumer goods and information, all these mainly

in African context, while in Latin America rural–urban linkages on the metropolitan fringe are not typically strong, and in Asia the increased mobility between sectors is blurring the economic distinctions.

iii. Conflictive land property ownership issues are typical on the fringe, because of pressures from squatters, private developers or speculators, large tenants, etc. Different land market conditions feature dual systems (informal and formal) and various property and tenancy arrangements such as rental or customary right systems.

iv. The demographic processes that underpin fringe development, where they can range from organized land invasion to planned expansion of the fringe subsuming existing rural villages, including the speculative subdivision of farmland near the city, the re-settlement of eradicated down-town slum inhabitants into public housing projects, or even temporary (daily, seasonal or circular) migration "especially in the Asian metro-fringe sites" (Browder et al. 1995).

1.3 Concept of Dynamic Urban Fringe Landscape

The perceived link between the city and the countryside evolved, shifting to new conceptual landscapes where rural–urban links were redefined. In this conceptual field, the urban fringe is considered as a transitional zone between city and countryside described as not a discrete area but rather a diffuse territory identified by combinations of features and phenomena, generated largely by activities within the urban zone (Nottingham and Liverpool Universities 1998). These areas are difficult to define as it is associated with problems inherent to the conceptualization of both rural and urban worlds.

A sharp distinction between urban and rural settlements generally assumes that the livelihood is based on two broad categories—agriculture based in rural areas and manufacture and service based in urban centre. Despite the activities are spatially separated, there is always a continued and varied exchange of resources between urban and rural areas. This sectoral interaction consists of primary activities in urban areas (like urban agriculture or dairying) or secondary and tertiary activities in rural areas (Tacoli 1998). This results in the boundaries between urban and rural areas increasingly blurred (Jones and Visaria 1997).

Lerner and Eakin (2010) used the word hybrid landscape for urban fringe which also incorporate significant changes in lifestyle, land use and livelihood. The fusion of livelihoods is seen all around the urban fringe and the younger generations commute to the urban locations to take advantage in their professional careers. The urban fringe demonstrates the potentiality to address the challenges of sustaining food security, environmental integrity and economic growth.

The conceptual background on urban fringe case studies by many such as McGee 1991; Robinson 1995; Lintelo et al. 2001; Kundu 2003; Dupont 2004; Iaquinta and Drescher 2001, Fazal 2013 etc. indicates that all studies have defined urban fringe according to their needs, scope of work and data availability.

1.4 Development of Urban Fringe in a Globalizing World

When analyzing the complex interactions between the global and the local, it is argued, there is a need to escape from "the idea of the global steam roller" constructing new regions, which is the common ideology underpinning globalization discourses. Almost 80 % of the world's inhabitants, particularly in Asia, Africa and Latin America, still exist in localities where networks remain primarily interpersonal and local; indeed, such local networks may be the crucial elements in their battle for survival (McGee 1997). McGee further argues that the globalization process is followed by an inevitable increase in urbanization and the emergence of global and sub-global systems of highly linked cities. If current patterns of urban concentration persist, the developing world is expected to experience the emergence of mega-urban regions as major components of their urban systems. Although it has been argued that the current phase of globalization in developing countries is the second after the one represented by their incorporation into the colonial system, the urban consequences are at least of the same importance. Delik et al. (2007) explained the uniqueness of peri-urbanisation in East-Asian countries in relation to its dependence on the metropolitan centres, capital accumulation and dynamic coexistence of urban and rural livelihoods.

One of the consequences of the first wave of "globalization" had been the creation of large primate cities dominating the urban hierarchies of their countries, such as New Delhi, Kolkata, Mumbai, Rio de Janeiro, Jakarta or Nairobi which were administrative centres and conduits for the flows of raw materials for the developing world. At the same time, dualistic societies were created, with much of the rural population living in poverty (as the informal sectors of the cities), dominated by a small elite of colonialists, foreign entrepreneurs and indigenous oligarchy (McGee 1997). Even if differences in the time and processes of decolonization (earlier in Latin America) have produced different patterns of urbanization—where Latin America has attained the levels of urbanization of developed countries, and Asia and Africa have just entered the "accelerated" phase of urban transition from rural to urban population —in each of these continents population from migration and endogenous growth of those large primate cities and industrial location on their fringes are actually expanding the urban areas to form what has been labelled extended metropolitan regions. McGee's major conclusion from this analysis is that macro-trends do not necessarily indicate a de-concentration of urban settlement or counter-urbanization.

In fact, while changing the scale and the definition of the urban agglomeration, these processes are actually occurring in a larger zone, where one can find at the same time residential outward movement, changing land use of the inner cores, industrial decentralization into new industrial states and the creation and amelioration of transportation networks. This larger zone is the Extended Metropolitan Region.

Following, McGee (1991), the features of
Southeast Asian EMRs are as follows:

- large and dense population engaged in wet rice cultivation;
- good transport networks;

- highly mobile population;
- an increase in non-farm (non-agricultural) activities;
- a mosaic of interlocking land uses;
- increased female participation in the labour force;
- lack of planning controls

Rigg has pointed out that the features cited above for many city-regions of the Asian countries being the same that one could find at the national level: "it could be said that in functional terms there is no "edge" to the EMR" (Rigg 1997 cited in Adell 1999). Rigg further maintains that: the formerly distinctive attributes of "rural" and "urban" are becoming increasingly blurred. In terms of physical fabric this can be seen in the expansion of Southeast Asia's metropolitan regions; at a functional level it can be seen in the spatial intrusion of industry into rural areas; and in human and household terms in the diversification of livelihoods and the increasing movement of people between regions and jobs. It is in interpreting these changes that the real challenge lies: what is driving the process and what are its implications, particularly for the people who are caught up in the maelstrom of change (Rigg 1997 cited in Adell 1999).

McGee (1991) having argued that these EMR are mainly shaped by the dynamic linkages between agriculture and non-agriculture, from the macro-regional point of view, an evolving system of urban linkages between these mega-urban regions is emerging as a phenomenon of growing economic interaction. A further consequence is the emergence of sub-global regions: regional blocs of states (such as the Asia-Pacific region) within the global system (McGee 1997).

1.5 Delimitation of Urban Fringe

Weherwein (1942), an American land economist at first define the rural–urban fringe as the area of transition between well recognized urban land uses and the area devoted to agriculture. Blizzard and Anderson (1952) defined it as areas of mixed urban and rural land uses between the point where full city services cease to be available and the point where agricultural land predominate. Ramachandran in 1989 criticized this definition as in Indian cities civic services are generally very poor and slums in Indian cities having hardly any amenities will have to be included within the fringe zone. Ramachandran defined rural–urban fringe in India as

> the rural–urban fringe is an area of mixed rural and urban populations and land uses, which begins at the point where agricultural land uses appear near the city and extends up to the point where the villages have distinct urban land uses or where some persons, at least, from the village community commute to daily for work or other purposes.

The rural–urban fringe limit of two metropolitan cities of India, namely Delhi and Bangalore was traced by following variables:

(a) Density of population—400 persons per sq. km or more
(b) Population growth in the preceding decade—40 % or more

(c) Females per thousand males—800 or less
(d) Proportion of workers in non-agricultural activities—50 % or more
(e) The outer limit of the city bus services or local train services.

But these criteria are not in fact satisfactory from conceptual angle as it has drawbacks like for West Bengal, where the density of 400 persons per sq. km is not applicable as the density of the state is much higher than that.

The delimitation of outer boundary of rural–urban fringe varies. In the UK and other European countries it can stretch out as far as six to ten miles from the urban edge, while in the US these distances may be twice or thrice as large (Audirac 1999). Thomas L. Daniels 1999 described the urban fringe as the area within 5–50 miles of a city that has scattered, low density development—fewer than 500 people per square mile—typically on 1–10 acre lots. In India the range is different depending on the availability and access of transportation and employment. The fringe zone extends from a minimum of 8 km to a maximum of 16 km around Delhi. In Bangalore the outer limits of the fringe extends up to a distance of 20 km from the centre of the city.

In the context of Indian cities, the outer limit of the urban fringe can be envisaged as the line at which the urban land use ends and the real village land use patterns, primarily agriculture, begin to appear. The inner limit of this zone of transition can be demarcated an area joining the points where any sort of non-urban or agricultural land use appear near the city. The criterion of daily linkages of people with the city is also important in determing the outer limit of the fringe zone in Indian context. The criterion of land use is as important as anywhere else. The fringe zones are characterized with great activity on part of the colonizers who purchase and sell land for residential and industrial uses. The western methods of demarcating fringe boundaries are not very much effective in Indian contexts, primarily because of different social and economic conditions. During the recent past some researchers have focused their attention to this issue. Nangia (1976) explaining the fringe zones as 'the zone which lies immediately outside the city, has strong interaction with the city in terms of daily communication, exchange of goods and services, and bears an urban reflection on the physical, occupational and demographic structure'. It is rural compared to the typical urban of the city and urban compared to the typical rural land scape.

Nangia has selected the following determinants to delimit this zone:

- Density of the population—isoline of 100 persons per square mile
- Population growth—relative change in population density—50 % growth
- Index of interactance—200 million person per mile
- Relative change in population density
- Females per one thousand males—800 females per one thousand males
- Workers engaged in non-agricultural activities to total workers—60 and 75 %

Besides these major determinants, many other criteria have also been employed in this study as spatial determinants like nearest neighbours distance, settlements along selected routes and linkages: and functional determinants like telephone exchange centers, postal delivery zones, frequency of bus services, shuttle services, milk chilling centers and railway marshalling yards. These methods of delimitation were employed to find a median boundary for the rural–urban fringe. Taking the nearest point of this boundary girdle to the city and the farthest point away from the city, two lines were drawn which form the inner and outer boundaries of the rural–urban fringe. Towards the city this zone merges into the city and towards outer limits, it graded into the peripheral rural areas.

1.6 Structure of Urban Fringe in India

The city and the surrounding areas comprise two types of administrative areas

(a) The municipal towns or nagar panchayats and
(b) Revenue villages or gram panchayats

The municipal towns tend to differ with distance from main city. Near the main city, the smaller municipal towns are part of the city geographically and have no identity of their own. These towns have the same level of municipal services as the main city. With distance from the main city the municipal towns are different as they have their own distinct identity, and a distinct set of problems relating to urban amenities and transportation. The rural–urban fringe may begin within the city limit, but this does not necessarily happen. The area beyond the metropolitan built-up area but contiguous to it having other municipal towns, census towns or fully urbanized villages constitutes the urban fringe as the part of rural–urban fringe.

The urban fringe is very much like the proper city with residential and commercial centres, but it generally lacks proper city services such as piped water supply, sewerage and garbage disposal facilities. Layout is haphazard and chaotic, streets are now widening and poorly maintained and there is proliferation of squatter settlements and slums. In parts, the urban fringe is well organized with posh residential colonies or industrial townships whose access roads are well maintained and have all the urban amenities. This, however, is an exception and not the rule.

1.7 Different Models of Urban Fringe

1.7.1 Pryor's Model

R.J. Pryor in 1968 distinguished the 'urban-rural' from 'rural-urban', in fringe area. He demarcate the urban–rural portion of the fringe is those areas having greater housing density than the average number of the total houses in the whole fringe area and a high proportion of non-agricultural land, higher rate of increase in population density, land use conversion and commuting (Fig. 1.1).

By contrast, the rural–urban portion is the rest of the fringe with the lower housing density than the average of the total dwellings and high proportion of farm

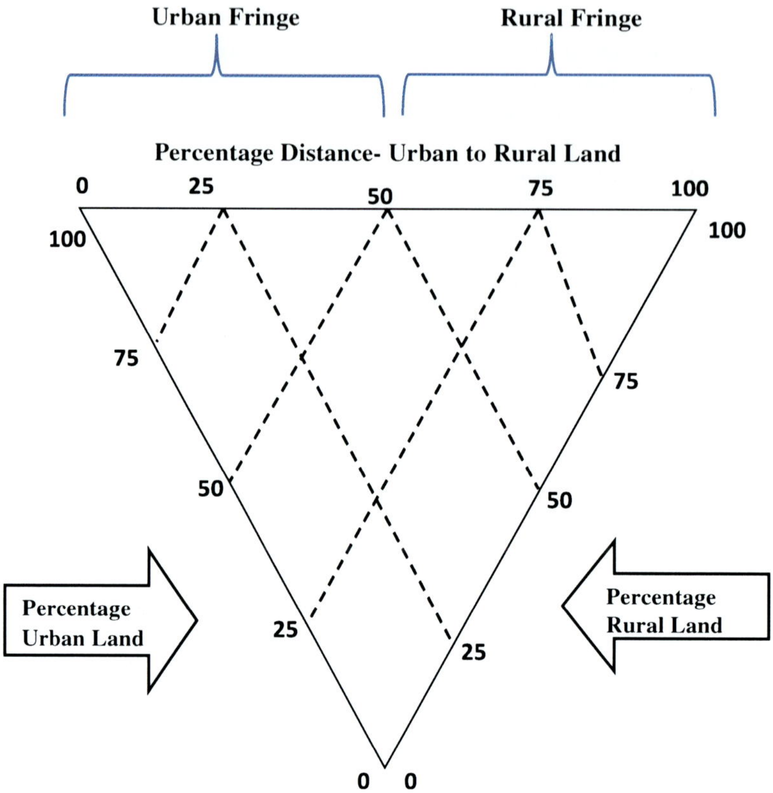

Fig. 1.1 Schematic diagram showing broad land use in Urban Fringe. *Source* Pryor 1968

land and lower rate of increase in population density, land use conversion and commuting (Audirac 1999).

1.7.2 Zonal Model of the Rural–Urban Fringe

Ilbery (1985) and Bryant et al. (1982) identify four zones comprising the rural–urban fringe cited in Audirac (1999)

(a) The inner fringe where the rural land is progressively converted to urban uses through the development process.
(b) Outer fringe surrounds the inner zone and contains the mixture characteristics of land use-rural and urban and urban uses are concentrated mainly along the highways and roads.
(c) The urban shadow is next to urban fringe and this zone is characterized by commuting patterns of part-time and hobby farmers and residents of small towns.

(d) Rural hinterland where the second home and recreational uses for urbanites are found amid extensive agricultural uses and open space.

1.7.3 The Bow Wave Model

In this model, four concentric zones are found which are based on agricultural activity. The zone closest to the urban centre concentrates the most intensive agriculture such as greenhouse production. The second band is dominated by the nursery production. The vegetable crop and dairy farming is practiced in third and fourth band respectively (Fig. 1.2).

1.7.4 The Ribbon Model

The ribbon model suggests that urban fringe development follows major roads linking the urban area to other cities and surrounding rural areas. Mabogunje (1990) argues that the construction of highways in "peri-urban and suburban districts has been an important mechanism for bringing more and more land within the sphere of the emerging capitalist mode of production". Improved roads enable the urban fringe poor to commute to work though the "inordinate distances" place a burden on them (Bryceson 2006). Kombe (2005) argues that improved roads attract middle-class residents and force the poor ever further away from highways. In addition, government-paid "car allowances" and increased access to automobiles permit residential developments along thoroughfares (Grant and Yankson 2003). In Dar es Salaam transportation improvements have resulted in a star-shaped urban development. Based on the ribbon model, it is expected urban fringe communities along highways to experience fast growth and high population density.

1.8 Delineation of Urban Fringe

As a landscape phenomenon, the urban fringe varies from city to city and from one time to another. Urban fringe that subzone of the rural–urban fringe in contact and contiguous with the central city, exhibiting a density of occupied dwellings higher than the median density of the total rural–urban fringe—a high proportion of residential, commercial industrial and vacant as distinct from farmland and a high rate of increase in population density, land use conversion and commuting (Pryor 1968).

From a review of some case studies of fringe areas, the following delineation techniques can be found (Tables 1.1, 1.2).

Table 1.1 Delineation of Urban Fringe

Definition	Delineation
1. Land use	• Specific, e.g. market gardens, Mixed, e.g. Between limits of exclusively urban and rural land • Valuation changes
2. Employment	• Census categories, e.g. RNF (Rural Non-Farming) • Community zone beyond central city boundary
3. Population density	• Rate of growth per year of inter censual
4. Social orientation	• Rural location, urban orientation of social activity
5. Transition, dynamism	• Undergoing change, e.g. increase in population density or vacant urban land.

Source Pryor (1968)

1.9 Rural–Urban Linkages

The physical boundary of urban built-up often extends beyond their administrative boundaries, so it is influence area. The influence area surrounding urban centers generally have major role in providing agricultural and other primary products for urban residents because of its location. It is rather difficult to make generalization on the nature of this region-urban fringe, which is depended on various interrelated as well as complex natural, social, economic and infrastructural factors existing within and even outside urban fringe. The rural–urban linkages reflect the regional development. Douglass (1998) in his qualitative research work on rural–urban linkages for policy research with reference to Indonesia showed the interdependencies between city and its rural hinterland provide potentialities for the development of both rural and urban development and gave the schematic diagram of five types of flows, i.e. people, production, commodities, capital and information between rural and urban areas. Bah et al. (2003) focused on the ways in which rural–urban linkages underpin and affect livelihoods in their case studies of Mali, Tanzania and Nigeria. The changes in the scale and nature of rural–urban linkages have an impact on livelihood transformation especially in agriculture sector. In the urban fringe, small farmers' access to land is subject to pressure and the urban residents invest in commercial farming. Poor physical infrastructure also affects the productivity and patterns and small farmers does not enjoy the facility of direct disposal at the urban market and the dealers or traders are acting the work as middle man between producer and retailer. Small-scale farming sector is negatively affected by severely limited access to credits, markets for agricultural producers and it becomes an unattractive option especially among younger generation.

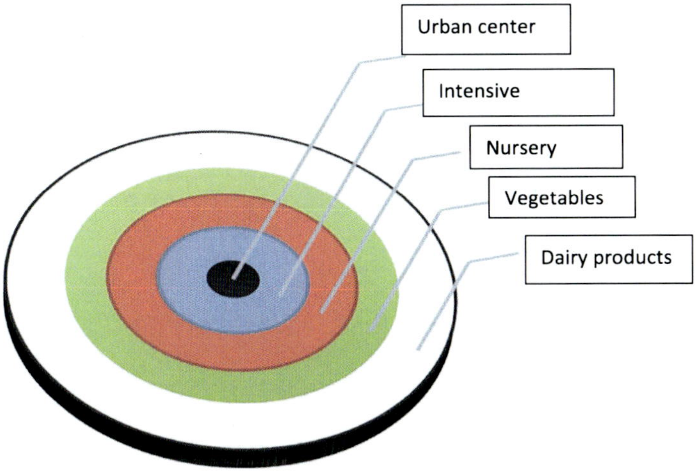

Fig. 1.2 The diagram of bow wave model. *Source* Hart 1991

Table 1.2 Rural-urban linkages/FlowsT

Rural structure/ structural changes

- Socio-economic structure/ relations
- Rural economy (sector)
- Rural production regimes
- Natural environment resources
- Infrastructure/ built environment

← →

People

- Labour commuting/ migration
- Othermigration(e.g. education)
- Shopping/ visiting / selling

Production

- Upstream linkages (inputs)
- Downstream linkages(processing, manufacturing)

Commodities

- Inputs
- Consumer non-durable/ durable
- Rural products

Capital

- Value added
- Savings/credit
- Migrant remittances

Information

- Production/sales/prices
- Welfare/social/political
- employment

← →

Urban functions/ roles

- Non-agricultural employment
- Urban services
- Production supplies
- Non-durable and durable goods
- Markets for selling rural products
- Processing/ manufacturing
- Information on employment, production, prices, welfare services

Source: Douglass 2007

1.10 Urban Fringe Linkages: The Changing Urban Fringe Linkages and Flows

Potter and Unwin (1989) analyzed the urban–rural relationships in developing countries under the general framework of the theory of dependence. Potter's article was marked by an ideological choice that proved an abstract socialist framework to analyze rural–urban linkages. It is hypothesized that the nature of the settlement pattern of a given country, is merely the reflection of deeper socio-economic inequalities. Thus, rather than trying to solve them through specific urban solutions as growth poles, new towns and the like, one has to turn to "more fundamental socio-political issues that are of crucial significance, not least, the mode of production and the development path that have been adopted in particular countries". (Potter and Unwin 1989)

Unwin (1989) focused from the point of view of the interactions, linkages and flows between rural and urban areas. These three terms being generally used in the literature interchangeably, Unwin proposed a framework which are to be seen as separate but closely linked concepts. At the broad theoretical level it suggests that there are economic, social, political and ideological linkages between urban and rural places. These find their physical expression in measurable flows of, for example, people, money and budgetary allocation. These flows are associated with interactions between people, places and objects, but do not in themselves actually embody those interactions (Unwin 1989).

Lerner and Eakin (2010) using the word hybrid landscape for urban fringe elaborated changing livelihood and their linkages with urban settlement. They stressed on urban fringe social changes which also incorporate significant changes in lifestyle, land uses and livelihood. The fusion of livelihoods in urban fringe is seen along streets, where livestock and gardens occupy back patios, and the younger generations commute to the locations of their professional careers. They want to show the potentiality of urban fringe to address the challenges of sustaining food security, environmental integrity and economic growth.

1.11 Urban Fringe Linkages: The City Expansion and Dispersal

The general move towards dispersal and location of growth on the peripheries or fringes of cities is becoming a world-wide phenomenon. Evidence of strong empirical regularities in the patterns of metropolitan development in market-oriented economies of both industrialized and developing countries (Ingram 1998). These movements towards the fringes are as follows:

- a dispersal process from the centre to the periphery of both population and employment, with the largest metropolitan areas converging to decentralized and multiple sub-centred areas, (cities in developing countries tend to have higher population densities, but the difference is narrowing).

- highly decentralized manufacturing employment and emerging specialization of the central business district in service employment.
- land-markets are strong determinants of this outward movement, land rents being closely related to development densities.
- urban hosing demand patterns are similar across cities in industrial and developing countries, but the supply side varies, as does the efficiency of the public infrastructures provision.
- increased reliance on road-based transport for both passengers and freight (industrial countries have experienced decreases in transit level as automobile ownership rises; developing countries have higher transit-ridership levels and a mix of options in terms of vehicle sizes and levels of services).

Ingram states that population growth in large cities usually promotes densification of less developed areas and expansion at the urban fringe, largely following either price constraints or preferences of households acting within the housing market. This kind of process has been analyzed in different countries. Indeed, the formation of vast and ever-expanding metropolitan regions seems to be "an inevitable feature of very large populous countries" in the developing world. The potential of growth of such agglomerations is been constantly enhanced by technological development, especially by major improvements in transports and telecommunications (Jones and Visaria 1997).

Adedayo in his study for developing countries also found that the urban frontier of housing estates and factories encroaches into formerly rural landscapes; these households in developing countries had become occupationally diversified, incorporating work patterns and networks tied into the metropolitan economy (Adedayo 2007).

In Southeast Asia, while suggesting the formation of an extended metropolitan region, Firman (1997) has described the fast process of land conversion in the Northern Region of West Java, Indonesia, mainly from agricultural to industrial and residential uses. This process has been led by domestic and foreign investment in the manufacturing, finance and service sectors, and has also been encouraged by a series of financial deregulation policies from the 1980s, aiming to stimulate economic growth.

The development of investment in housing and industrial estates has thus transformed mainly peripheral agricultural areas to large town and industrial estates. An influx of population into these peripheral areas, mainly middle-income groups, has boosted land speculation and a strong and dynamic activity of developers, not always under the regulation of the state (Fazal 2001). This process must also be interpreted in the light of a general economic and physical restructuring of the region, where Jakarta City, the core, is shifting its former industrial function to a business, finance and services centre. Meanwhile, population growth in peripheral areas is far exceeding the one in the core (Firman 1997).

Puntip (2009) also stated that the metropolitan region of Bangkok as a whole was in the stage of suburbanization indicating by the high rate of population growth in urban fringe areas, by transforming economic base to non-farming economy, and by rapid expansion of built-up areas.

The mega-urban regions resulting from processes like those mentioned above are often composed of one or more large urban cores linked by major transportation axes plus the periurban zones and an extensive zone of mixed rural–urban land use alongside the main routes (McGee 1991). Common features of these rural–urban areas include the following:

- rapid commercialization of agriculture;
- expansion of transport systems;
- an employment shift from farming to other activities, accompanied by migration or commutation to cities. Factories and other non-agricultural activities are also developing within the zone but outside the city proper.

Another characteristic feature of these mega-urban regions is that the distinctions between urban and rural are tending to become increasingly blurred as a result of technological changes and globalization processes that make available to rural dwellers, goods and services that were the preserve of urban residents only a very short time ago (Jones and Visaria 1997).

1.12 Urban Fringe Linkages: The Secondary Cities and Urban Diffusion

The work of Rondinelli (1983, 1984), although not explicitly, could be seen as trying to conceal two approaches, one based on the central place theory (and thus close to the growth centreapproach) and one more rural oriented, that could be paralleled with the agropolitan approach proposed by Douglass and Friedmann (1998). Rondinelli states that the inequitable development is not related to the urban population, but with the city size system and hierarchy of the country. A geographically dispersed pattern of investment, based in a network of secondary cities and market towns or rural villages, is seen as important to achieve widespread development (Unwin 1989). Rondinelli (1983) considers that increasing poverty in rural areas have played a determinant role in a certain "neglect of cities". This effect has been reinforced by the evidence of the disparities in living conditions and conditions of development between urban and rural areas in the developing world.

Rondinelli (1983) also stated that lack of a system of secondary cities that supports economic activities and decentralized administrative functions which are necessary for the development of both urban and rural sectors. At the same time, rural–urban linkages are the promoters of this widespread development. He mentioned these linkages are crucial because the major markets for agricultural surpluses are in urban centres, most agricultural inputs come from organizations in cities, workers seek employment in towns as technological developments frees rural labour, and many of the social, health, educational, and other services that satisfy basic human needs in rural areas are distributed from urban centres (ibid). At the same time, he argued that no matter how carefully development goals are conceived it only matter as to how these linkages worked, as well as how they are conceived and by whom. They cannot be achieved in isolation from the cities or entirely through "bottom-up" stimuli. Economic growth with social equity requires both accelerated agricultural development and expansion of urban industry and commerce. (ibid.)

Unwin (1989) stresses the importance of Rondinelli's contribution to the consideration of the urban–rural interaction in that it relies heavily on "linkages both between rural areas and small cities, and on those between smaller and larger cities". He also argues that Rondinelli's approach to development planning can be summarized in two main points: "it is essentially concerned with the implementation of change through the manipulation of the urban settlement hierarchy, and it is implicitly based on a 'free market', capitalist framework of change" (Unwin 1989).

1.13 Linkages and Livelihood in Urban Fringe

The urban fringe linkages have a bearing on the livelihood and well-being over space and can create resilient regional economies (Fazal 2014). There can be various types of linkages, here an attempt has been made to give an explanatory description of existing linkages within the urban fringe. The key interest in describing linkages in urban fringe is to understand the nature and direction of these flows and to identify driving forces to promote mutually supportive rural–urban development (Table 1.3).

1.13.1 Demographic Flows

Rural to urban migration is a central characteristic of urbanization in India as in many developing countries. Migration is one of the few avenues open to poor households to increase income and combat increasing improvishment. The dominant trend is for members of a rural household to migrate to urban areas to sell their labour and remit their income to their respective families back in rural areas. With the opening up of economies, diversifying activities have created opportunities leading to rural–urban migration in the recent past (Sjoberg 1994; Liu 1991; Wong 1994 and Fazal 2013). Migration also has observable gender biases depending upon the type of jobs available in rural and urban areas and social environment of the rural place in terms of facilitating or hindering male/female migration. In India, men have traditionally dominated migrant streams but some newly industrialized countries with labour intensive trans-national assembly line operations have also observed women as dominant migrants to urban areas (Lynch 2005). In either situation, migration is a highly selective process within households and among rural as well as urban places of origin and destination.

Table 1.3 Livelihoods and the Rural–Urban Continuum

Rural ≫		≪ Urban
Livelihoods drawn from crop cultivation, livestock, forestry or fishing (i.e. key for livelihood is access to natural capital) Access to natural capital as the key asset and basis for livelihood Access to land for housing and building materials not generally a problem More distant from government as regulator and provider of services Access to infrastructure and services limited (largely because of distance, low density and limited capacity to pay?) Fewer opportunities for earning cash; more for self-provisioning. Greater reliance on favorable weather conditions.	Rural-urban fringe	Livelihoods drawn from labor markets within non-agricultural production or making/selling goods or services Greater reliance on house as an economic resource (space for production, access to income-earning opportunities; asset and income-earner for owners—including de facto owners) Access to land for housing very difficult; housing and land-markets highly commercialized More vulnerable to 'bad' governance Access to infrastructure and services difficult for low-income groups because of high prices, illegal nature of their homes (for many) and poor governance Greater reliance on cash for access to food, water, sanitation, employment, garbage disposal…
Urban characteristics in rural locations (e.g. prosperous tourist areas, mining areas, areas with high value crops and many local multiplier links, rural areas with diverse non-agricultural production and strong links to cities….)		Rural characteristics in urban location (urban agriculture, 'village' enclaves, access to land for housing through non-monetary traditional forms…)

Source Satterthwaite and Tacoli 2003

In India, migration has several dimensions. It is not a one time, one way movement of people, but rather, it is part of a dynamically adaptive strategy that allows poor households to tap into resources available in both sending and receiving urban and rural areas in order to cope with conditions of poverty and to try to enhance their wellbeing. Especially, north Indian migrants as a survival strategy return back to their rural native place during harvesting period to augment their earnings from the urban employment. An estimated 20 million people migrate temporarily each year in India, of which almost 2/3 migration is rural to urban.

The pattern of migration may vary in different regions according to their socio-economic-cultural setup but migration is certainly pervasive and dynamic process of poverty alleviation. In India, remittances account for about 1/3 of annual incomes of poor households in rural areas (Deshingkar 2006). Migrant remittance to rural households plays an important role in well-being of their rural-based families. Remittances at household level make a significant contribution in raising education, health and living standard in rural areas. It also makes a difference in rural society as a whole as investments also flow into the development of rural farming enterprises promoting rural economic development. These investments from migrants in social welfare, education, housing etc. in their native rural areas fetch them social respect and recognition.

1.13.1.1 Driving Forces for Migration

Empirical studies and historical trends indicate that rural–urban migration has been primarily driven by differences in expected income over space, which in the extreme is manifested in the inability of households to earn-incomes to sustain their existence. Several factors contribute to both income inequalities over space and the fall below levels of subsistence in particular regions. Traditional approaches to migration have relied on the notion of "push-pull" factors are the main explanatory elements. Individuals rationally decide to migrate because they are attracted by the bright lights of the city which promise, in long term to offer better economic opportunities than the countryside.

1.13.1.2 Structural Changes in Economies and Migration

Economic policies and government intervention in the economy greatly influence what will be produced and consumed and in turn profoundly affect rural–urban migration. They affect the structure of rural production and the availability of jobs to both the urban and rural workers. Migrants are rational decision makers that respond to income opportunities and resource availability (Tacoli 1998). India is undergoing structural adjustments and the effects of these policies have bearing on rural–urban migration also. For instance, small farmers in the absence of government subsidies and support, found it difficult to overcome high input costs. They also cannot afford to hold on to their produce and sell at higher prices. They are also vulnerable to price squeeze as the cost of inputs and consumer goods rose much steeper than the prices they fetch from primary commodities. This pushes small farmers to urban areas for sustenance. Empirical evidences suggest that movements are not necessarily permanent but many times circular and seasonal as migrants try to capture the economic benefits of both rural and urban areas.

1.13.1.3 Environmental Considerations for Rural Resources and Migration

It is increasingly being recognized that a significant driving force behind rural to urban migration is the lack of access by poor households to natural resources. Empirical studies show that there is a strong economic relationship between direct access to natural resources and the sustainability of livelihood in rural areas. Poor households depend greatly on environmental natural resources to meet their minimum needs. Like harvesting forest materials, fishing or collecting fuel wood etc., which supplements their farming or off farming employment. Therefore, to the rural household income, this is part of a 'portfolio' of earnings contributing to total household earnings, which is both subsistence and income as well. But in recent past governmental policies have greatly restricted access and exploitation of these resources. The conservation policies which protect rural environmental natural resources by restricting local communities from accessing subsistence resources and changes in land tenure associated with certain types of rural development and conservation investments prevent access to environmental resources and inadvertently create poverty which stimulates migration.

1.13.1.4 Conflict, Instability and Migration

The increasing globalization of the development process also affects natural socio-economic and political changes even in the remote rural settlement. This instability makes the marginalized rural population difficult to guarantee the conditions for survival. This also leads to migration for protecting themselves as well as sustaining their living. Famine, riots, land acquisition are some example for migration instigated by regional or even local instability and conflicts.

1.13.1.5 Migration and Gender

The causes, nature and outcomes of urbanization are found to be highly gendered. Gender selective migration reflects a range of social and economic imperatives and constraints surrounding men's and women's activities, power and status in rural communities. It also induces new patterns of behaviour, access to resources and social organization (UNDP 1996). Migration tends to be male dominated in India, because women's chances of getting job are lower than livelihood possibilities in rural areas. With income being the important motive for migration, it is recognized that women may not move to urban areas under their own volition, but at the behest or through pressure from their rural families (UNDP 1996). Traditionally, women are responsible to take care of the general well-being of other household members than their male counterparts. In some cases their migration is resisted so that remittances continue to flow for the benefit of her extended family.

1.13.2 Flow of Materials and Goods

Exchange of goods between urban and rural areas are an essential element of rural-urban linkages, and the most recent generation of spatial policies considers market interactions as a crucial factor in the development of rural areas, reflecting the global trend towards market-led strategies (Tacoli 1998).

Raw materials derived from the extraction of natural resources in rural areas tend to flow towards urban centres where most of the industrial transformation and manufacturing are based. Rapid urbanization draws enormous amounts of resources into urban centres from its hinterland. The "virtuous circle" model of rural–urban, development creation of non-farm jobs and the diversification of urban activities, especially in small towns close to areas of agricultural production absorbs surplus rural labour, raises demand for rural produce and once again, boosts agricultural productivity and rural incomes (Evans,1990; Evans and Ngau 1991 and UNDP/UNCHS 1995).Theoretically, rural links to urban markets supposed to promote the purchase of key inputs for rural agriculture from adjacent urban areas, which would spur the productivity and income in rural areas. This may lead to increases in purchasing power of rural population creating demand for manufactured urban commodities. However, the flows of goods from urban to rural areas have not been up to the expectations. On the contrary, rural resources are often depleted in the urban expansion and development process. This in many ways depletes the resource base for rural production and fails to generate local economic linkages beyond the extractive process. In such situation, rural and urban regions are unable to build on the synergies that have been expected of them. There is an urgent need to link towns and cities with rural production in a more positive way. This would gradually offer opportunities to improve living conditions in rural areas, the very reason for which migration occurs.

Majority of rural producers are unable to participate in ever increasing urban demands due to poor collection, storage, marketing and transportation services. The institutional support is also found to be far deficient than required. Lack of credit to consistently produce at sufficient scales needed to supply urban markets or the role of middleman that discriminate against or ignore the potential of smaller, more distant producers. Generally, the access to urban markets is controlled by a few buyers; spatial monopolies and dominance over rural producers dampen economic returns to farmers and inhibit further processing of resources with the region.

1.13.3 The Flow of Capital

Financial flows are an important component of rural–urban interaction. There is a diverse range of ways in which capital flows between rural and urban areas occurs which can go in either direction. While there are diverse capital flows such as remittances to rural areas and urban based investments in rural areas. The dominant pattern in such flow is generally concentrated and attracted to few prosperous urban centres. An integral part of this rural to urban transfer is the concentration of resources in hands of wealthy and powerful individuals of the society. This phenomenon signified by the growing inequality and income gap between rich and poor as well as rural and urban dwellers.

1.13.4 The Flow of Waste

The flow of wastes between urban and rural areas is closely related to both the flows of raw materials derived from natural resources and extracting the benefits from the processing of these materials. Rural areas are the source of raw materials for almost all manufacturing and processing industries and are in advertently the base for the generation of capital. The dominant rural–urban pattern of natural resource use and waste is such that natural resources in rural areas are developed and channelled to urban centres where they are consumed and the wastes are returned back to rural areas, commonly termed as ‘ecological footprints of the cities’.

The pattern and magnitude of flows in resources and wastes is at the core of several problems that are identified by the international community as global environmental problems, as shrinking biodiversity and forest loss etc. urban production and consumption is a significant driving force for rural resource degradation. Both production and consumption are fed by the exploitation of resources and with the rapid acceleration of production and consumption patterns, it is widely accepted we are catapulting towards environmental breakdown.

The links between rural–urban flows of waste is circular and mostly have regional consequences. The cyclic nature of rural–urban environmental linkages is illustrated by the relationship between unsustainable livelihoods in rural and urban areas. Unsustainable levels of demand that tends to be concentrated in urban centres encourage inappropriate development and the degradation of rural environments.

1.13.5 The Flow of Information and Ideas

The flows of information between rural and urban places have a core function in effective development planning. Decision about the flow of materials, resources and capital flow to the rural areas with pronounced urban bias, while the ability for rural voices are somewhat limited. The poor human development of rural areas is largely attributed to poor communication linkages. The decisions about their economies and welfare from higher levels of administration are firstly made ignoring the local ground condition; secondly these decisions are filtered in such a way that benefits of these decisions are further reduced.

The developments of information and communication technologies have helped significantly in connecting rural and urban areas more effectively and at low costs. While flows of goods and people are tangible and observable, the flows of information and ideas that usually accompany them are far more complex and difficult to research. This is partly because it is less tangible as well as the relations that structure the flows of other elements also influence the flows of ideas.

The movement of people to and from cities and the numerical growth of city dwellers have had profound effects on societies more generally. In particular, increased movement of people has resulted in increase of information.

The information flow between rural and urban areas falls in four categories

- The information flow on resources from rural areas
- The information flow about markets, price of agricultural and other rural commodities

- The information flow manifested through political decision making
- The information flow on opportunities and technologies.

The spread of mass media has helped considerably in blurring the differences in urban fringe. The people have greater access to newspaper, radio, television, phones etc. which is facilitated by improvements in technology. Despite traditionally authoritarian state control, these mediums have showed in breaking the state control, increasing free and increased exchange of information and ideas.

Simon (2008) addresses the complexities of changing urban fringe production and livelihood systems in the context of rapid urbanization in poorer countries, distinctive urban fringe challenges of appropriate and flexible planning and development, and the future prospects for enhanced sustainability in the most challenging category of development-environment interfaces.

1.14 Conceptual Framework

The present work on urban fringe livelihoods will be explored with focus on agricultural production and transformation of livelihoods from primary sector to secondary and tertiary sector and also the livelihood diversification. All the issues mention will be put under the conditions and trends of the study site. As the study area located on the edge of the city, the impact of urban process including increasing demand, growing market will be assessed (Fig. 1.3). Accessibility is another important indicator for the analysis of livelihood strategies in the urban fringe areas. Different zones have different access to different livelihood resources. Further their living conditions and quality of life evaluated and the importance of city proximity and economic transformations was also assessed.

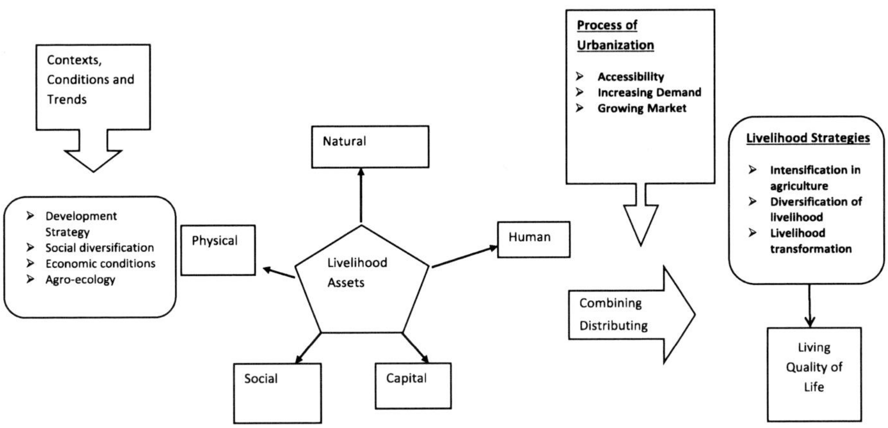

Fig. 1.3 Conceptual Framework of the Study

References

Adedayo A (2007) Socio-spatial transformations and the urban fringe landscape in developing countries. Presented Paper at United Nation University Institute For Environment And Human Security (UNU-UHS) Summer Academy On Social Vulnerability and Resilience Building In Mega city. Munich, Germany

Adell G (1999) Theories and models of the peri-urban interface: a changing conceptual landscape. Draft Report, Development Planning Unit, University College, London. http://www.ucl.ac.uk/dpuprojects/drivers_urb_change/urb_economy/pdf_Urban_Rural/DPU_PUI_Adell_THEORIES_MODELS.pdf

Audirac I (1999) Unsettled views about the fringe: rural-urban or urban-rural frontiers?

Bah M et al (2003) Changing rural–urban linkages in Mali, Nigeria and Tanzania. Env&Urbanization, 15 (1): 13–24

Blizzard SW Anderson WF (1952) Problems in rural-urban fringe research: conceptualization and delineation. Pennsylvania State College, School of Agriculture, Agricultural Experiment Station

Browder J, Bohland JR, Scarpadi JL (1995) Patterns of development on the metropolitan fringe—urban fringe expansion in bangkok, jakarta and santiago. J of the Am PlanAsso 61 (3)

Bryant CR, Russwurm LH, McLellan AG (1982) The City's Countryside: Land and its Management in the Rural-Urban Fringe. Longman, London

Bryceson DF (2006) Growing out of spatial poverty: Growth, sub-national equity and poverty reduction policies—a five-country comparison. A Synthesis Paper Prepared for DFID Rural-Urban Change Team

Daniels TL (1999) When city and country collide: managing growth in the metropolitan fringe. Island Press, Washington D.C

Delik H et al (2007) Peri-urbanisation in East Asia: a new challenge for planning? International Development Planning Review 29(4). http://www.academia.edu/1880342/periurbanisation_in_East_Asia_a_new_challenge_for_planning

Deshingkar P (2006) Internal migration, poverty and development in Asia. Asia 2015: Promoting Growth, Ending Poverty. London

Douglass M (1998) A Regional network strategy for reciprocal rural-urban linkages. In: Third World Planning Review on Agenda for Policy Research with Reference to Indonesia. 20(1)

Douglass M, Friedmann J (eds) (1998) Cities for citizens: planning and the rise of civil society in a global age. John Wiley, London

Dupont V (1997) Les 'Rurbains' de Delhi, Espace, Populations. Societes 2(3): Paris, ORSTOM, 225–240

Dupont V (2004) Urban development and population distribution in Delhi: implications for categorizing population. In: Champion T, Hugo G (eds) New Forms of Urbanization: Beyond the urban-rural dichotomy. Ashgate, Aldershot, pp 171–190

Evans HE (1990) Rural-urban Linkages and Structural Transformation. Report INU 71, Infrastructure and Urban Development Department, The World Bank, Washington DC

Evans HE, Ngau P (1991) Rural- Urban Relations, Household Income Diversification and Agricultural Productivity. Development and Change 22:519–545

Fazal S (2000) Urban expansion and loss of agricultural land. A GIS Based Study of Saharanpur City, India. Environ and Urbanization 12(2):133–149

Fazal S (2001) Application of remote sensing and gis techniques in urban sprawl and land use mapping. A Case Study of A Growing Urban Centre in India. Asian Profile 29 (1), Canada, 45–62

Fazal S (2013) Land use dynamics in a developing economy. regional perspective from India. Springer Publication, The Netherlands

Fazal S (2014) Peri-urban livelihoods: opportunities and challenges. Concept Publication, New Delhi

Firman T (1997) Land conversion and urban development in the northern region of west Java. Indonesia. Urban Studies 34(7):1027–1046

Gant R, Robinson G, Fazal S (2011) Land-use change in the 'edgelands': policies and pressures in london's rural–urban fringe. Land Use Policy 28 (1): Elsevier Science Publication, United Kingdom, 266–279

Grant R, Yankson P (2003) City profile: Accra. Cities 20(1):65–74

Hill R (1986) Land use change on the urban fringe. Nat and Res 22:24–33

Ingram GK (1998) Patterns of metropolitan development: What Have We Learned? Urban Studies 35(7)

Iaquinta D, Drescher WA (2001) More than spatial fringe: an application of the peri–urban typology to planning and management of natural resources.In: International conference on Rural-Urban Encounters: Managing the Environment of Peri-Urban Interface, Development Planning Unit, University College, London, Unpublished paper presented at DPU on November 2001

Jhonson JH (ed) (1974) Suburban Growth: Geographical Processes at the Edge of Western City. John Wiley and Sons

Jones Gavin W, Visaria P (1997) Urbanization in Large Developing Countries: China, Indonesia, Brazil, and India. Clarendon Press, Oxford, London

Kabra KN (1980) Developing an Urban Fringe, Blueprint for Voluntary Action. Indian Institute of Public Administration, New Delhi

Kombe WJ (2005) Land Use Dynamics in Peri-Urban Areas and Their Implications on the Urban Growth and form: The case of Dar es Salaam, Tanzania. Habitat International 29:113–135

Kumar MS (2001) Devouring Agglomerations and their Outgrowths in India: Beyond Rural—Urban Divide. Unpublished paper presented at DPU International Conference on Rural-Urban Encounters: Managing the Environment of Peri-Urban Interface. Development Planning Unit, University College London, London

Kundu A (2003) Urbanisation and Urban Governance: Search for a Perspective beyond Neo-Liberalism. Economic & Political Weekly, 38(29)

Leeming L, Soussan J (1979) Structures at the Fringe of the City. Int Soc Sci J 31(2):273–281

Lerner AM, Eakin H (2010) An obsolete dichotomy? Rethinking the rural–urban interface in terms of food security and production in the global south. The Geograph J Royal Geograph Soc (with The Institute of British Geographers)

Lintelo DT, Marshall F, Bhupal DS (2001) Delhi Peri-Urban Agriculture. Retrieve from: fao.org/DOCERP

Lucas P, Van Oort GMRA (1993) Dynamics in a rural-urban fringe: working and living on the edge of the city of Utrecht. Faculteit Ruimtelijke Wetenschappen, Rijksuniversiteit Utrecht, Utrecht, Nederlandse Geografische Studies 154

Liu APL (1991) Economic Reform, Mobility Strategies, and National Integration in China. Asian Surv 31(5):393–409

Lynch K (2005) Rural-urban interaction in the developing world rural-urban interaction in the developing world. routledge perspectives on development. Routledge, London

Mabogunje A (1990) Urban Planning and the Post-Colonial State in Africa: A Research Overview. Afr Stud Rev 33(2):121–203

McGee T (1991) The emergence of desakota regions in asia: expanding a hypothesis. In Ginsburg N, Koppel B, McGee T (eds) The Extended Metropolis: Settlement Transition in Asia, University of Hawaii Press, Honolulu

McGee T (1997) Globalisation, Urbanisation and the Emergence of Sub-Global Regions. In: Watters RF, McGee T (eds) Asia Pacific: New Geographies of the Pacific Rim. Hurst and Company, London

Nangia S (1976) Delhi Metropolitan Region—A Study in Settlement Geography. K.B. Publishers, Delhi

Nottingham and Liverpool Universities (1998) Literature review on peri-urban natural resource conceptualization and management approaches, initial draft, peri-urban production systems. research, Nat Res Sys Prog, DFID, London

Pryor RJ (1968) Defining the Rural-Urban Fringe. Soc Forces 47:202–215

Puntip J (2009) Urbanization and Changing Settlement Patterns in Peri-urban Bangkok. Kasetsart J. (Soc. Sci) 30: 303–312
Rakodi C (1998) Review of the Poverty Relevance of the Peri-urban Interface Production System Research, Report for the DFID Natural resources Systems Research Programme, 2nd Draft
Ramachandran R (1989) Urbanisation and Urban Systems in India. Delhi: Oxford University Press (Reprint 2001)
Rao MSA (1991) A rural community on the Delhi metropolitan fringe. In: Rao MSA Bhat C Kadekar LM (eds) A Reader in Urban Sociology, Orient Longman, New Delhi, 389–409
Robinson I (1995) Emerging Spatial Patterns in ASEAN Mega—Urban Regions: AlternativeStrategies. In: McGee T, Robinson I (eds) The Mega Urban Regions of Southeast Asia. UBC Press, Vancouver, pp 78–108
Rondinelli Dennis A (1983) Secondary Cities in Developing Countries: Policies for Diffusing Urbanisation. Sage Publications, Beverly Hills
Rondinelli Dennis A (1984) Applied Methods of Regional Analysis: the Spatial Dimensions of Development Policy. Westview Press, Boulder, Colorado
Saini NS (1989) Rural Development at Metropolitan Fringe: Resource Conservation Approach. Anmol Publications, New Delhi
Satterthwaite D, Tacoli C (2003) The Urban Part of Rural Development:The role of small and intermediate urban centres in rural and regional development and poverty reduction. 9226IIED Report. Human Settlement Working Paper
Schenk H (1997) The Rurban Fringe: A Central Area between Region and City: The Case of Bangalore, India. In: Chatterji M, Kaizhang Y (eds) Regional Science in Developing countries. MacMillan, New York, pp 212–223
Siddhartha K, Mukherjee S (1997) Cities Urbanization and Urban Systems. Kisalaya Publications Pvt. Limited, New Delhi
Simon D (2008) Urban Environments: Issues on the Peri-Urban Fringe. Annual Review Environment Resource 33:167–185. http://arjournals.annualreviews/org/doi/pdf/
Sjoberg O (1994) Rural Retention in Albania: Administrative Restrictions on Urban Bound Migration. East European Quarterly. 28(2):205–234
Smith TL (1937) The Population of Louisiana: Its Composition and Changes. Louisiana Bulletin 293:24
Swindell K (1988) Agrarian Change and Peri-Urban Fringes in Tropical Africa. In: Rimmer D (ed) Rural Transformation in Tropical Africa. Ohio University Press, Ohio, pp 98–115
Tacoli C (1998) Rural-Urban Interactions: A Guide to the Literature. Environment and Urbanization 10(1)
Thakur B (1991) Land Utilization and Urbanization, the Rural-Urban Fringe. Deep & Deep Publishing House, New Delhi
UNDP (1996) Urban agriculture: food, jobs and sustainable cities. New York: United Nations Development Program, Publication Series for Habitat II, Volume One
UNDP/UNCHS (Habitat) (1995) Rural-Urban Linkages: Policy Guidelines for Rural Development. Paper prepared for the Twenty-Third Meeting of the ACC Sub-Committee on Rural Development, UNESCO Headquarters is 31 May–2 June 1995
Potter R, Unwin T (eds) (1989) The Geography of Urban-Rural Interaction in Developing Countries: Essays for Alan B. Mountjoy. Routledge, London
Unwin T (1989) Urban-Rural Interaction in Developing Countries: A Theoretical Perspective. In Alan B, Mountjoy. Potter R Unwin T (eds) The Geography of Urban-rural Interaction in Developing Countries: Essays Routledge, London
Weherwein EG (1942) The Rural Urban Fringe. Economic Geography 18:217–228
Wong L (1994) China's Urban Migrants—The Public Policy Challenge. Pacific Affairs 67(3): 335–356
Yadav CS (1987) Rural-Urban Fringe. Concept Publishing House, New Delhi

Chapter 2
The Study Area: Aligarh Urban Fringe

2.1 Location of Aligarh City

The district of Aligarh extends from 27° 29′ to 28° 11′ north latitude and 77° 29′ to 78° 38′ east longitude (Fig. 2.1). River Ganga and Yamuna make the east and west boundary, respectively, while in north, Bulandshahar district, and in south, Hatharas district are situated. Aligarh district has been divided into five tehsils, such as Koil, Autrauli, Sikandrarao, Iglas and Khair. Aligarh city is located in koil tehsil of lodha block and the city lies almost in the centre of Aligarh district. Aligarh city is a class I city of the state of Uttar Pradesh, situated in its western part between Ganga and Yamuna rivers at 27° 53′ north latitude and 78° 35′ east longitudes. It is the head quarter of Aligarh district, spread over an area of about 44.82 km^2. Presently, Aligarh city has been divided into 70 wards. This city has gained importance because of its proximity to the national capital (New Delhi) and also known for the Aligarh Muslim University.

2.2 Historical Evolution of Aligarh City

In early history, the settlement was associated with the kingdom of *Surasena*. The settlement in this region began around 1500 B.C. An ancient town existed here which was known as *Koil* or *Kol,* named after an *Asura* king. Atkinson regards origin of Aligarh as a Buddhist settlement following 400 B.C. Around second century A.D., the area came under the possessions of *Mauryas, Sakas, Kushans* and *Naga* rulers. Archaeological remains of fifth to ninth century A.D. belonging to *Gupta* period and *Harsha* empire have also recorded from Aligarh. In tenth century A.D., one of the old settlements identified in Aligarh city as *Achal Tal* (tank) area located in south-east. Twelfth century A.D. marks the arrival of Muslims in this region. In 1194, *Qutubuddin Aibek* took the fortress of *Kol and Balbon* and constructed *minar* tower in 1253 just close to present *Jama Masjid* on the great mound. Four dynasties such as the

N. Banu and S. Fazal, *Livelihood and Wellbeing in the Urban Fringe*,
The Urban Book Series, DOI 10.1007/978-3-319-39660-6_2

Slave's from 1194 to 1240, *Khilji's* from 1290 to 1320, *Tughlaq's* from 1320 to 1414 and *Lodha's* from 1451 to 1526 contributed to the control and construction of Aligarh. During the reign of *Ibrahim Lodhi* (1526 A.D.), a permanent fortress known as *Ramgarh* was constructed, which is still existing. *Umar Khan* built the fort of *Mohammadgarh* in 1526, which was later known as Aligarh. During the mediaeval period, *BalaiQila*area acted as the socio-economic and political hub. Throughout this period, there were walls and gates around the city, i.e. *Madar, Delhi, Turkman, Sasni* and *Aligarh Darwaza.* In the eighteenth century, *Sabit Khan* was appointed as the governor of *Kol* and he constructed the *Jama Masjid.* Marathas took over the Aligarh fort in 1785. Thereafter the Britishers rule the city, during their rule the city spread northwards and got economically progressed. After the British occupation in 1804, the present district of Aligarh was formed and the term Kol ruined. With the increase of population industries were developed. Initially cotton industry (1802) was established, besides this, other industries like crude glass, glass bangles and later lock industry were established and flourished.

The most remarkable during the evolution and development of the city was the establishment of *Mohammedan Anglo Oriental (MAO) College* in 1875 which is now known as Aligarh Muslim University. This was founded by *Sir Sayed Ahmed Khan* to educate Muslim people. Now, it is one of the central universities of India and is situated in civil line area.

Aligarh city because of its historical background and evolution consists of three different sociocultural areas.

- The ancient area, dominated by Hindu Population
- The mediaeval area, dominated by Muslim Population
- The modern area, dominated by mixed Population of both Hindus and Muslims

These areas instead of losing their identity with time have retained it. Incidences of communal riots have strengthened the segregation on communal lines.

2.3 Administrative Set up of Aligarh City

Aligarh city is the main urban and administrative centre of Aligarh district. Aligarh came into existence as a district in 1804. The district is further divided into five sub districts (tehsils), each with its own divisional headquarters. These five tehsils have been further sub divided into 12 blocks (Fig. 2.2). These blocks are again subdivided into total 1210 villages.

Aligarh city is located at the Lodha block of Koil tehsil. Municipal Corporation was formed for Aligarh city in 1975, while the formation of Aligarh Development Authority took place in 1981. The First Master Plan and Second Master Plan were drafted in 1981 and 2001, respectively, for the city. The city which initially had 19 wards (1971), but now it is divided in 70 wards (2008) and a mayor is elected to administer the city.

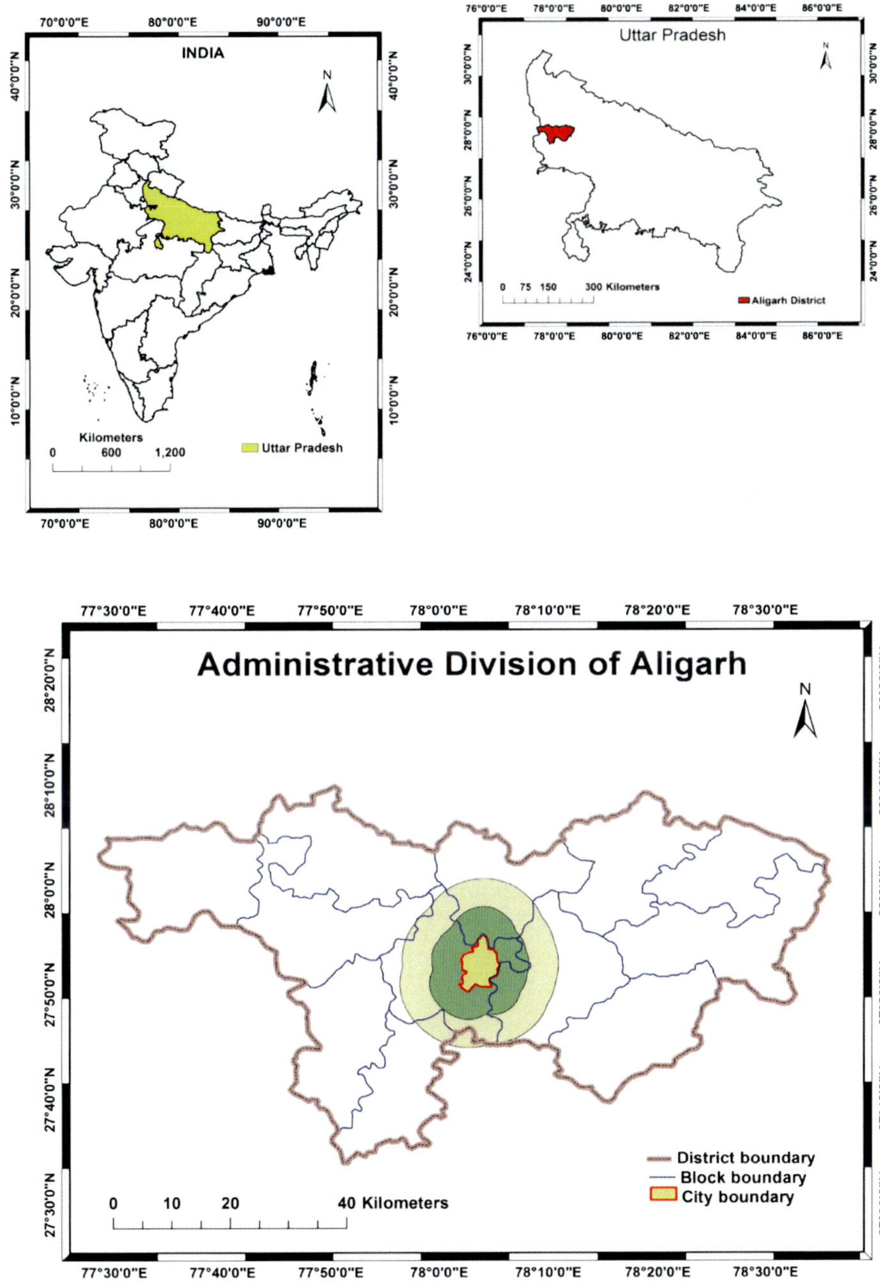

Fig. 2.1 Location Map of Aligarh City

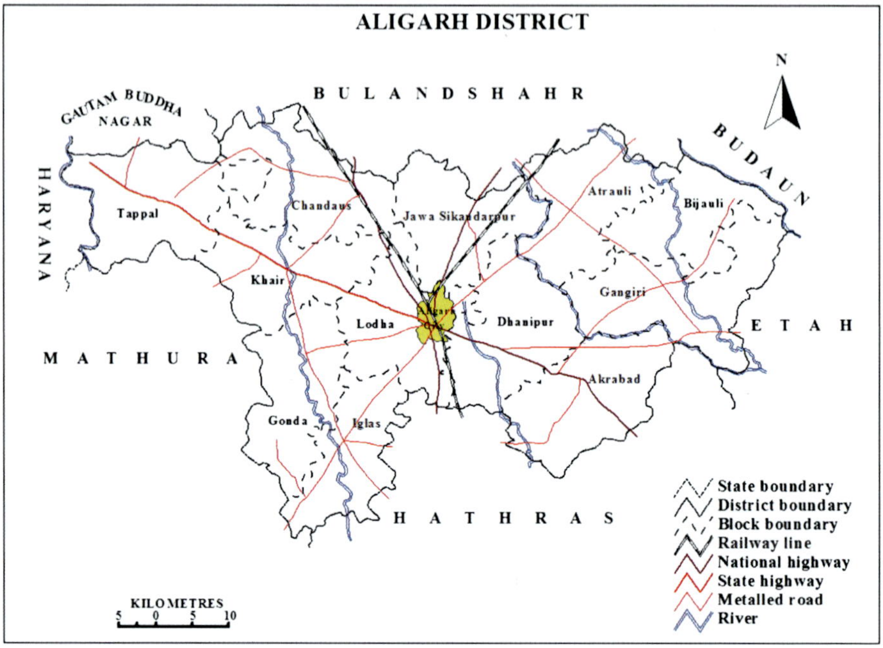

Fig. 2.2 Map of administrative division of Aligarh

2.4 Population Growth in Aligarh City

The total population of Aligarh city is 872,575 according to 2011 census. Although the decadal growth rate of city population is shown to be fluctuating (Fig. 2.3 and Table 2.1) but the share of the city population to district total population has been gradually increasing since 1971 (Fig. 2.4).The city experienced a high growth rate of 49.75 % in the years 1981–1991 (Fig. 2.3) due to immigration from nearby

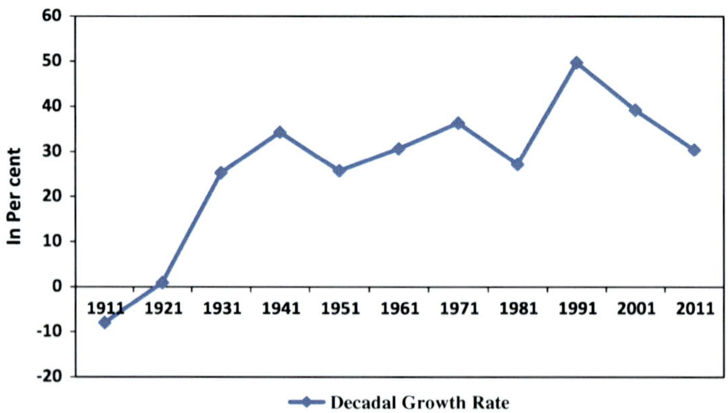

Fig. 2.3 Decadal growth rate of Aligarh city population. *Source* Population Census of India

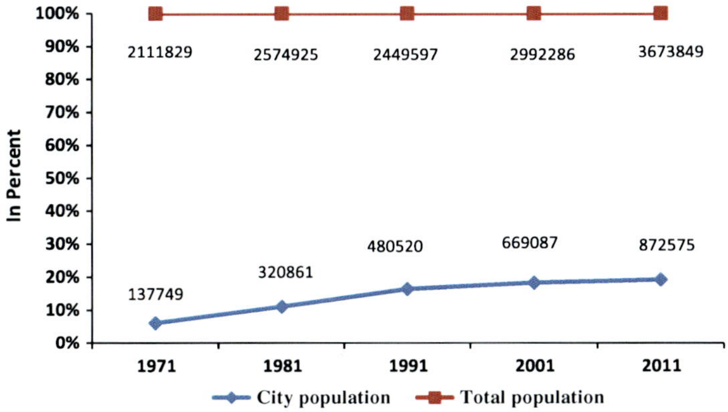

Fig. 2.4 Share of city population to the district population. *Source* Population Census of India

Table 2.1 Decadal growth of population of Aligarh City

Census year	Population	Difference of population between two decade	Difference in per cent
1901	72,084		
1911	66,344	−5740	−8
1921	66,963	619	0.93
1931	83,878	16,915	25.26
1941	112,655	2877	34.31
1951	141,618	28,963	25.71
1961	185,020	43,402	30.65
1971	252,314	67,294	36.36
1981	320,861	68,547	27.17
1991	480,520	159,659	49.76
2001	669,087	188,567	39.24
2011	872,575	203,488	30.41

Source Population census of India

villages. The years 1991–2001 also have a high growth rate of 39.24 % due to the expansion of Aligarh Nagar Nigam boundary limits in the year 1994 in which rural population is also included. With the increase of population, the city requires developments in housing, amenities and facilities and employment opportunities, etc. which as in other cities of India do not match with the city population. This leads to stress on existing facilities creating problems such as mushrooming of slums, environmental pollution, unemployment, congestion, water and electricity crisis, etc.

Table 2.2 Growth of marketing centre in the Aligarh City

Year	Total number of marketing centre	Decadal growth rate
1971	10	–
1981	21	110.00
1991	35	66.67
2001	48	37.14
2011	82	70.83

Source Deputy Regional Marketing Officer, Aligarh

2.5 Marketing Centres in Aligarh City

Aligarh city has a good number of markets to meet the requirements of the people. The CBD of Aligarh encompasses Railway road, *Phaphala, Barahdwari and Chauraha Abdul Karim*. There is an agglomeration of both wholesale and retail trade in this area. Certain areas of the city are famous for wholesale and retail goods. The important ones are *Mamu-bhanja* area for hardware items and electronics goods, *Phaphala* for pharmaceuticals items, *Bans ki Mandi* for timber and house building materials and *Dhanipur* for vegetables and cereals and it is one of the eight new agricultural markets in Aligarh district, out of which, three are located on the periphery of the city. Beside Dhanipur, Sarsol fruit market and Harduaganj grain and vegetable market are also located at the urban fringe areas of Aligarh city. There are about 90 marketing centres in Aligarh city (Table 2.2). Every locality in the city has a market centre and some of them have grown up in recent years to become new commercial hubs such as *Jamalpur, Dodhpur, Amir Nishan, Kelanagar, Kuwarsi, Soot mill, Sasni gate*, etc.

2.6 Status of Infrastructure Facilities in Aligarh City

There are ten major roads which all converge at Gandhi park, almost at the centre of the city. These roads include national highway (N.H.-91) or G.T. Road which connects the city with *Delhi and Kanpur*, N.H.-93, which connects with *Anupshahar* (Fig 2.5). Apart from these roads, there are several state highways which radiate out of the city to connect other important towns of the state. Among them, the important are *Ramghat* road, *Khair* road, *Gonda* road, *Agra* road and *Mathura* road, etc. These roads have acted as the important agent helping in the development of the city. Recently big factories and other industrial establishment have sprung up along these roads providing the city a radial dimension. Besides these roads, there is well-developed internal road network connecting important

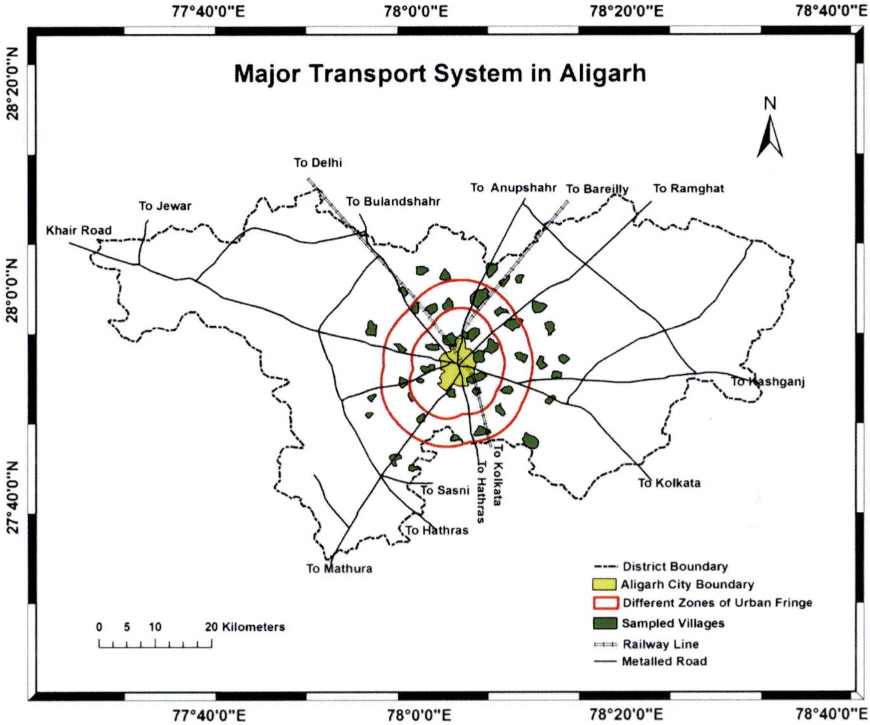

Fig. 2.5 Map of the major transport system in Aligarh

locations within the city. However, the roads of older city are narrow and congested while the newer part has wider roads.

There are two main bus terminals—*Masoodabad* bus terminal along G.T. Road and *Gandhi* park bus Terminal from where U.P. State Road Transport Corporation buses serve cities all over the state of Uttar Pradesh and many other cities in Uttarakhand, Rajasthan, Madhya Pradesh, Delhi, Punjab and Haryana. There are five other private bus terminals in Aligarh city, which connect the city with nearby rural settlements (Table 2.3).

Aligarh is also connected with rail network, the Aligarh railway station is a railway junction. It is on the main railway route connecting national capital Delhi with states of West Bengal, Orissa, Bihar, Jharkhand, North-Eastern states of Assam and most of the parts of U.P.

Table 2.3 Passengers movements by bus from different terminals in Aligarh City

S. no.	Name of the bus terminals	No. of buses	Average no. of passengers	Name of the route
1	Gandhi Park Terminal	176	8800	Lucknow, Agra, Moradabad, Mathura
2	Masoodabad Bus Terminal	75	3375	Delhi
3	Khair	40	1600	Zebar, Tappal, Khair
4	Gonda	24	760	Gonda
5	Chharra	25	1025	Chharra, Sankar
6	Gangiri	20	800	Gangiri
7	Jalai	35	1400	Jalali
8	Autaruli	33	1485	Autrauli, Ramghat, Bijauli
9	QasimpurRamghat	20	800	Sadhuasram
10	Anupshahar	90	3150	Aniupshahar, Nonora
11	Sambhal	22	550	Sambhal
12	Pahasu	12	370	Pahasu
13	Barauli	12	510	Barauli
14	Amrauli, Barauli, Pahasu	24	960	Amrauli, Barauli, Pahasu
15	Qasimpur	20	600	Qasimpur
Total		628	26185	

Source Aligarh Development Authority, 2001

2.7 Employment Prospects in Aligarh City

The city offers different types of occupational opportunities, which fit into different sectors and draw different incomes. The city has well-developed secondary activities where a large number of skilled and semi-skilled workers (37 % of total workers) are engaged. There are a total of 5506 industrial units in 2011 while, in 1981 there were only 104 industries in Aligarh city (Tables 2.4 and 2.5). Of these, there are 3500 small-scale industries, 2000 medium-scale and 6 large industries (Table 2.5). There are three kinds of major industries namely food processing, lock industry and household industry. Most of the industries are located in the residential areas within the city limits.

The most famous is lock industry besides it other industries like engineering industries, building fitting, electrical goods industries, chemical industries have also helped the city for its development (Table 2.5). A good number of small-scale units of brass work, edible oil, sugar, cotton textile, cotton ginning, dairy, glassware, readymade garments, handloom and weaving provide jobs to the city population. The city also has a large number of people engaged in tertiary sector (61 % of total workers), associated with Aligarh Muslim University, other educational institutions, banks, post offices, health, communication, trade and commerce, etc.

Table 2.4 Growth of industries in Aligarh City (1981–2011)

Year	Small-scale industries		Large-scale industries		Total number of industries	Decadal growth rate
	Number of units	Number of workers	Number of units	Number of workers		
1981	95	585	9	956	104	–
1991	439	2851	11	2165	450	332.69
2001	3316	9641	13	3985	3329	639.78
2011	5650	34,132	14	5165	5664	70.14

Source Industrial Directory, DIC, Aligarh, 2011

Table 2.5 Types of industries in Aligarh City

Name and type of industries	Number of large-scale industries	Number of medium scale industries	Number of small-scale industries	Total
Food processing	6	0	0	6
Lock industry	0	2000	0	2000
Household industry	0	0	3000	3000
Other	0	0	500	500
Total	6	2000	3500	5506

Source Aligarh Nagar Nigam, 2010

2.8 Nature-Based Activities and Its Economic Input

The study area receives scanty and erratic rainfall, where 80 % of total rainfall happens in the month of south-west monsoon from mid-June to September. Rest of the year it is mostly dry. On the other hand, soil is generally sandy loam having high rate of permeability. Thus, the agriculture is heavily depended on the irrigation especially through tube well and canal irrigation which are expensive. The urban shadow effect is greatly affecting the agriculture activity in the surrounding rural areas of Aligarh city. There is only 2 % workers associated with agriculture and its share to the total working force is gradually decreasing over the year while people's dependency is gradually increasing on tertiary economic activities in the city (Fig. 2.6). The land value is increasing and the high cost of farming inputs have made small and marginal farmers vulnerable in modern monetized urban economy. This has seen as one of the reasons for the trend of transformation from farming activities to non-farming activities in the study area. Another cause of livelihood transformation is the high growth of population which is resulting fragmented and smaller land holdings. All this is making agriculture, which was preferred livelihood option as lesser favoured occupation (Fig. 2.6 and Table 2.6).

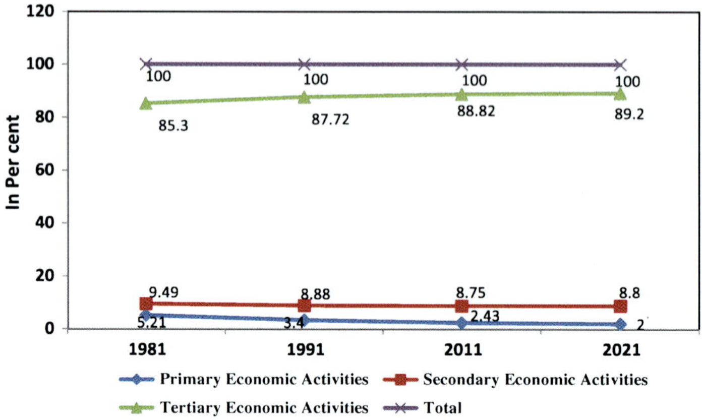

Fig. 2.6 Working population of Aligarh city engaged in different sectors of economic activities. *Note* The data for the year 2021 is projected by Aligarh development authority (Aligarh Master Plan 2001–2021). *Source* Population census of India

Table 2.6 Working population engaged in different economic activities in Aligarh City

Economic activities	1981		1991		2011		2021 (projected)	
	Workers	%age	Workers	%age	Workers	%age	Workers	%age
Primary workers and agricultural labourers	3320	5.21	4234	3.40	5648	2.43	7104	2.00
Households workers	4253	6.68	6883	5.51	12,416	5.34	19,536	5.50
Other than households industries	17,615	27.65	40,449	32.38	78,976	33.98	121,778	34.20
Construction	1790	2.81	4206	3.37	7919	3.41	11,722	3.30
Trade and commerce	13200	20.72	29,304	23.46	55,412	23.85	83,472	23.50
Transportation and communication	7270	11.41	7400	5.92	11,510	4.95	19,536	5.50
Other activities	16,250	25.52	32,432	25.96	60,496	26.03	92,352	26.00
Total workers	63,698	100	124,908	100	232,377	100	355,200	100
Total population	257,065	24.78	490,202	25.48	872,575	26.63	118,4000	30.00

Source Aligarh master plan, 2001–2012

2.9 Development of Urban Fringe Around Aligarh City

Rural areas surrounding the Aligarh city lack the amenities and facilities like employment in secondary and tertiary sectors, education, transportation communication system, water supply, etc. which are available in the city. These availabilities are the pull factors to attract migrants from the surrounding villages. The influx of population through migration into the city caused the residential population to grow and the city expanded physically to accommodate it, first through the development of vacant land within the city itself, and later by the slow encroachment on land in areas lying outside the city limits. The net result has been the intrusion of urban land uses within the rural areas surrounding the rapidly growing Aligarh city.

There is a particular pattern of stages through which village community (surrounding Aligarh city) passes as the villages get transformed from a rural entity to an urban one. Due to the increase in population by natural growth and migration in Aligarh city, it expanded from city centre towards surrounding villages to accommodate these people. The village economy gets tied up with the growing city. There is phenomenal increase in linkages with daily movement from the village to the city in order to supply the different commodities such as perishables, like vegetables, milk, and flowers or low valued natural resources such as bricks, earth, fuel wood, etc. to the city. There are metal roads connecting the peripheral rural settlements with Aligarh city and even the city transport system also extends its services to these settlements. Daily visits to the city job, domestic requirements, recreation and medical facilities have become increasingly important. With time and improvements in the communication and transport facilities, the linkage with the city has improved significantly. The village in the fringe area is no longer inhabited by traditional village population only. It also becomes a living place for industrial workers, etc. who like to take advantage of comparatively low rents. Thus, the rural settlements around the Aligarh city have evolved into rural–urban fringe. The city spreads in Lodha, Akrabad, Sasni, Dhanipur, Jawan, Atrauli, Iglas, Gonda and Khair block. Many of these areas, till recent past were predominantly rural settlements but now absorbed the urban character also. Many villages have come into the direct influence of urban expansion and now lie in the urban limit. Kishanpur, Dodhpur, Begpur, Jamalpur, Bhamola and NaglaBaraula are all examples of urban absorption as these settlements witnessed in mute silence their transformation from rural to urban.

Chapter 3
Urban Fringe: The Features of Livelihood

The urban fringe is described as a complex zone spatially located between urban and rural areas where livelihoods of the population are under constant pressure from urban forces. The nature of the urban fringe varies considerably according to the level of urbanization, the economy and the geographical location from urban centre. It is a common understanding that urban growth has not only impacts over natural and built environment but also livelihoods of people living in the urban fringe. Urbanization offers livelihood choices for people in the urban fringe and also enables them of better access to urban services and infrastructure. This provides opportunities for them to exploit benefits from both urban and rural activities. In this sense, urban fringe is symbiotically interlinked with urban and rural areas. Yet, urbanization is also accompanied by intense competition for land, tenure insecurity, population pressure, stressed infrastructure, and polluted environment. The urbanization processes in many cases have posed threat to traditional livelihood practices. These situations and the monetized economy in urban fringe have significantly affected the natural resource-based primary economic activity and also result in the poverty and vulnerability to the livelihoods of poor people.

This chapter seeks to present the result and discuss the findings gathered through the data of field survey and the Population Census Data over last three decades. The analysis of surveyed data has been done at household level as well as at individual level. The result as presented in this chapter hold the factual evidence with interpretation or evaluations in relation to theory, existing literature work and other relatable fact of the study.

N. Banu and S. Fazal, *Livelihood and Wellbeing in the Urban Fringe*,
The Urban Book Series, DOI 10.1007/978-3-319-39660-6_3

3.1 Livelihoods in Urban Fringe

The urban fringe is one of the most critical acumen. It is not the extent of the resource base contained in the urban fringe that makes it critical to society, but rather the fact that the land and resources therein are subject to competing, often conflicting, demands. The urban fringe has been placed under pressure increasingly with increased demands due to rapidly changing social, cultural and economic values associated with land and its resources. Changes in main livelihood activity over generations or even within a generation often are adaptations to changing circumstances. In the urban fringe, these circumstances involve diminishing natural resource bases including land, direct and indirect opportunities presented by new infrastructure, changes in social structure (Simon et al. 2006 cited in Adu-Ampong et al. 2008). The rapid sprawl of cities, outside its political boundaries and beyond its real physical–cultural domain possesses several attendant conflicts and stress and has also stimulated various adaptations of economic activities.

3.2 Categorization of Livelihood

Livelihood in Aligarh's urban fringe can be categorized into two broad categories

- Non cash based livelihood
- Cash based livelihood

3.2.1 Non Cash Based Livelihood

The households engaged in non-cash-based livelihood in urban fringe are engaged with food production, fuel or fodder collection from common land. But this form of livelihood has been almost wiped out in Aligarh's urban fringe. It is evident from the survey that only four individuals reported to be engaged in non-cash-based activities (Table 3.1). Beside the dependency on common land, another type of non-cash-based livelihood is manual scavenging of village roads. In the urban fringe of Aligarh city, especially women from poor families are forced into this

Table 3.1 Aligarh urban fringe: cash based and non cash based livelihoods of workers among sampled households

Zone		Livelihoods						Total
		Non-cash based	Cash based					
		Non cash	Casual employment		Tertiary	Farming	Dairying	
			Labouring in agricultural sector	Labouring in non-agricultural sector				
Immediate urban fringe	Count	**0**	**0**	**117**	**392**	**148**	**173**	**830**
	% within zone	0.0 %	0.0 %	14.1 %	47.2 %	17.8 %	20.8 %	100.0 %
	% within fringe	0.0 %	0.0 %	38.9 %	61.4 %	37.5 %	50.3 %	48.7 %
Intermediate urban fringe	Count	**2**	**8**	**85**	**156**	**119**	**87**	**457**
	% within zone	0.4 %	1.8 %	18.6 %	34.1 %	26.0 %	19.0 %	100.0 %
	% within fringe	50.0 %	38.1 %	28.2 %	24.5 %	30.1 %	25.3 %	26.8 %
Distant urban fringe	Count	**2**	**13**	**99**	**90**	**128**	**84**	**416**
	% within zone	0.5 %	3.1 %	23.8 %	21.6 %	30.8 %	20.2 %	100.0 %
	% within fringe	50.0 %	61.9 %	32.9 %	14.1 %	32.4 %	24.4 %	24.4 %
Total	Count	**4**	**21**	**301**	**638**	**395**	**344**	**1703**
	% within zone	0.2 %	1.2 %	17.7 %	37.5 %	23.2 %	20.2 %	100.0 %
	% within fringe	100.0 %	100.0 %	100.0 %	100.0 %	100.0 %	100.0 %	100.0 %

Notes
i. Figures in bold are absolute number of sampled respondents in the particular distance zones
ii. Figures in underlined are percentages to the total from that particular distance zone
iii. Figures below the underlined figures are percentages to the total of that particular variable
Source Based on field survey (2013)

job.[1] They receive miserable food and clothes in return (City Sanitation Plan, Ministry of Urban Development, Government of India 2009). In Aligarh's urban fringe, the urban elements linked to the growing importance of inclusion with in a monetized economy while the rural element was linked to depleting natural resources. Poor people lacked control over these changes in urban fringe which adversely affecting their non-cash-based livelihood in particular.

Aligarh Urban Fringe: Evidences of Non Cash Based Livelihood

She is a street cleaner by profession. After sweep the village roads, she collects food from households.

[1]Those who are engaged in this type of activities in the urban fringe of Aligarh city, are known as Bhangi, Valmiki or Dhanuk. These are all local term. They sweep their village roads and clean toilets then collects food from households.

Livelihood depends on fuel collection from common land.

This was most glaring around the natural resources, destruction of natural and common resources (common land, forest and water bodies) to urban space demands, loss of soil fertility (due to extraction of construction materials, dumping of garbage) insecurity of tenure (due to opportunistic farming practices) and soil and water pollution (due to urban wastes).

The pressure of growing populations made natural common resources as the most contested asset in the urban fringe. Political empowerment at grass root level has also led to encroach upon the common natural resources. At some places the common resources are used for the benefit of community, while at other places they are used for the benefit of their own clan or caste. But in most cases and places, general deterioration and disappearance of common land is observed. All this led to reduction in the availability of the natural resource-based products needed to support traditional non-cash-based livelihood elements. The households which were dependent on these common lands for their livelihood have now either earn livelihood from labouring in rural areas or they have migrated to city where they sell labour for cash returns. Here, the sample survey found that women (old, widow, abandoned by family) are the main sufferers, who now cannot labour to earn with disappearing common land and they are at the mercy of government aid (*vradhavastha pension*).

3.2.2 Cash-Based Livelihoods

Cash-based urban fringe livelihoods can be categorized in four broad categories

- Casual employment
- Business and trade
- Farming activity
- Dairying activity

3.2.2 (i): Cash-Based Casual Employment

The casual employment in Aligarh's urban fringe has many dimensions. Most of the lesser educated and lesser skilled workers belong to this category. There is ever increasing demands for this low level, mostly labour oriented jobs in urban fringe. The employment opportunity of this kind is typically casual, short term and seasonal in nature fetching low and varying wages.

This kind of labour work can be divided into two categories

3.2.2 (i) a: Labouring in Agricultural Sector

This type of activity is becoming lesser popular in the urban fringe of Aligarh city where only 1.2 % out of the total workers are engaged themselves in agricultural field (Table 3.1). Agricultural labourers are not to be found in the immediate urban fringe and it is found in intermediate and distant urban fringe but the share is very less. Now the people do not prefer this type of activities, as in this type of employment the wages are comparatively lesser. This is despite the competition for labour apparently driving agricultural wage levels higher. This kind of labouring in the Aligarh's urban fringe has a peculiar feature, there is a considerable shortage of labour in the agricultural sector but its remuneration does not increase the levels of labour in nonagricultural sector of urban areas.

Another feature of this kind of labouring is that it is highly seasonal and is not gender biased. In fact, women are found to have almost equal opportunity for work. Even in some cases, the wages for women are higher than their male counterparts. Another feature of this kind of livelihood is that male members during peak demand (farming seasons—especially sowing or harvesting) come back from urban centres and join this sector for limited period, till they are required and afterwards they return back to the urban centre. This seasonal migration to them is like "*Paid holiday*". As they take leave from work in urban areas, join their families and also earn. They also take advantage of this period for managing social events during their stay with the family.

This type of labour in the agricultural sector is highly seasonal, the main reason for which is that the male members of the households work outside agricultural sector for sustaining their livelihoods.

3.2.2 (i) b: Labouring Outside Agricultural Sector

This is a cash-based livelihood opportunity mostly in unorganized sector, a common feature of urban fringe. In this sector, 18 % of the total work force is engaged as labourers outside agricultural sector (Table 3.1). It is adopted by households belonging to rural poor, uneducated, unskilled landless or small uneconomic land holdings, week and down trodden section of the society. Traditionally, these labourers worked in the rural agricultural sector, many of them were exploited with poor wages. The urban fringe offered them opportunities to sell their labour for remittance in cash.

This provided an opening for rural labourers to move away from primary sector and earn relatively higher wages, but these kinds of employment also have some inherent drawbacks. These are typically casual, seasonal, short term 'by day' employment. The urban proximity is an important requirement for this kind of employment. Another feature of this employment is that it is highly gender biased with male dominance.

A great variety of activities draw labourers to work for secondary and tertiary sector in urban fringe. Importantly, although they get associated with higher level of economic activity (secondary and tertiary), they do not possess any skill or expertise and only sell their labour. They work in factories as labourers or porters, construction workers, quarrying, brick making, rickshaw pulling, rag picking, watchman, domestic helpers, scavengers, cleaners, etc. Among all these activities, factory worker is most preferred because of regularity of work while construction works fetch best wages.

These activities also have specific spatial existence. The urban fringe of Aligarh has dominance of different pockets of activity in different locations. Like industrial area, where industrial labourers are required, university and civil lines area require domestic workers, rickshaw pullers, watchmen, etc. while Ramghat, Agra and Mathura road, etc. (mostly newer part of the city) have demand for construction workers.

Another feature which was observed in this cash-based employment outside agricultural activity was that labourers were found to make frequent changes to their work. Interestingly, these changes were not mainly for better wages (as there was no big difference) but for easier less labour oriented work.

3.2.2 (ii): Cash-Based off-Farm Livelihood in Tertiary Sector

It is the most popular economic activity in the Aligarh's urban fringe because 38 % out of the total sampled workers are engaged in this sector. One striking feature observed in Aligarh's urban fringe was declining preference to agricultural activity as livelihood option among younger generation. This change had a clear urban fringe character, becoming particularly strong with proximity to the city, where people also have a greater choice of livelihood options. Nearly, about half of the workers in immediate urban fringe maintain their livelihood from this livelihood option and with increasing distance from the city it is decreasing. The people were also found to make more and frequent changes to their livelihoods in immediate urban fringe than distant urban fringe.

The immediate urban fringe had reported a wide and diverse range of income generating activities, mostly requiring some skills (most of them are craft-based, artisanal activities or providing lower level services) and found to be employed by some organized group or self-employed. The people belonging to this section were relatively younger, have attained some level of education but were distancing themselves from labour-oriented agricultural activities. This younger urban fringe population prefer to work and stay in urban area, enjoy the advantages of urban infrastructure and perform a wide range of activities, which require some skills, such as mason, carpenter, driver, tailor, welder, plumber, electrician, hair dresser, auto mechanic, etc. If they do not qualify for any of these activities, then they opt for small time trading which is also one of the livelihood options for this section. The urban proximity is essential for these activities because they are required only in urban areas.

The most common examples of trade by men were the trading in a retail shop. Artisan work consisted of activities typically carried out by men. Activities such as carpentry, bricklaying and the general household repair shops are other common livelihood options in this category.

3.3 Livelihoods Trends

In the rapidly changing economic scenario, diversification of economic activity is the process by which households try to sustain themselves and improve their standards of living. In the study area, there are evidences of transformation and diversification of economic activities. Economic development and diversification of economy are the main reasons leading to this change. Significantly, this diversification of economy is said to be more pronounced in urban fringe. Urban forces including proximity to city can create new job opportunities for urban fringe households (Table 3.2).

Census data were tabulated to find the status of economic activities in Aligarh's urban fringe. The census data signalize towards the trend where working force in the primary sector has been decreasing on the other hand there has been a steady growth of tertiary sector. In 1981, on an average, the share of work force in primary sector was 77 % while it decreased to 28 % in 2011 (Table 3.3). Workforce engaged in tertiary sector was 16 % in 1981 and it increased to 67 % in 2011 (Fig. 3.1). The results corroborate the trend that there is a gradual mutation of work force from primary to secondary and tertiary sector. Moreover, this trend was more pronounced in immediate urban fringe (Figs. 3.1, 3.2, 3.3 and 3.4).

Table 3.2 Differences in the characteristics of the poor in urban, urban fringe and rural areas

Livelihoods	Poor in urban	Poor in urban fringe	Poor in rural
	Poor are concentrated in the informal economy, trades and as domestic labourers.	Poor are involved in both agricultural and in the urban informal sector. The very poor in the urban fringe are typically agricultural labourers.	Poor are involved mainly in agricultural work.
Nature of and reasons for migration	The urban poor do not tend to migrate back to villages. When they migrate it is permanent and often to other cities in search of urban jobs such as the construction industry.	The urban fringe poor are involved in circular migration commuting back and forth from cities to villages for shorter period from daily to weekly commuting.	The rural poor migrate due to distress factors, sometimes seasonally, sometimes permanently.
Type of wages	Wages are paid in cash.	Wages are paid in cash for labour.	Wages are paid in cash and/or in kind.
Opportunities	There are more opportunities to find work throughout the year.	Some of the poor have the opportunity to work both in agriculture or take advantage of urban and urban fringe opportunities. Others can not avail of urban opportunities due to lack of skills, health, alcoholism, lesser mobility, particularly the females. This group also suffers the negative consequences of urbanization in the form of insufficient well paid work in the urban fringe.	Work is agriculturally related and completely season bound. The majority does not work in the off-season periods, but a small minority engages in small enterprises.
Human capital	The poor have varied skills relevant to petty business and jobs in the informal sector.	The poor who seek urban opportunities may not be as skilled as their urban counterparts. The urban fringe youth try to pick up urban skills or skills related to agriculture which they can encash in the cities (gardening).	The poor have skilled relevant to agriculture.
Government programmes	Government programmes are designed to target the urban poor. However, the access is limited and depends on local influential people.	Villages in the urban fringe have a mixed scenario. Those under the Municipal Corporation have no access to rural government programmes or subsidized credit, but some urban facilities are provided. They enjoy only partial benefits compared to the truly urban population but also have to pay higher taxes than rural villages. Villages under the Panchayati Raj System have access to all the rural development programmes.	Government programmes cater to the needs of rural poor. Where in the recent past developments have taken place. The governmental efforts as well as local population awareness have contributed for this change.

Table 3.3 Aligarh urban fringe: economic sector wise workers in sampled villages

Sampled villages	Primary sector				Secondary sector				Tertiary sector			
	1981	1991	2001	2011	1981	1991	2001	2011	1981	1991	2001	2011
Immediate Urban Fringe	–	–	–	–	–	–	–	–	–	–	–	–
Maheshpur	80.23	57.94	44.91	12.95	0.00	10.32	0.6	2.73	19.77	31.75	54.49	84.32
Barhati	82.97	68.8	59	35.86	2.47	13.57	4.67	2.25	14.56	17.64	36.33	61.89
Wajidpur Nada	74.45	28.14	49.22	33.28	2.46	40.42	12.85	5.05	23.10	31.44	37.93	61.67
Kwarasi	76.49	35.37	8.41	11.65	12.35	21.92	2.64	2.81	11.16	42.71	88.94	85.54
Talaspur Kalan	76.54	35.4	22.52	23.24	5.14	3.75	2.35	6.43	18.32	60.85	75.13	70.33
Ramgarh Panjoopur	50.95	40.97	17.31	8.37	27.58	23.54	4	3.28	21.47	35.49	78.7	88.35
Dhanipur	45.39	35.09	9.05	5.60	15.36	33.54	6.54	5.98	39.25	31.37	84.41	88.41
Kasba Kol	71.73	43.85	13.9	18.55	5.24	28.12	9.22	4.40	23.04	28.02	76.87	77.05
Alinagar	42.16	83.33	10.5	50.63	23.53	10.42	2.5	9.49	34.31	6.25	87	39.87
Sarsaul	54.66	41.97	12.6	6.26	17.88	26.49	6.87	4.22	27.46	31.54	80.53	89.52
Haivatpur Fagoi	71.17	47.24	59.35	32.33	0.00	33.86	1.94	1.88	28.83	18.9	38.71	65.79
Average	**63.95**	**41.94**	**18.04**	**12.81**	**11.56**	**23.58**	**5.62**	**4.31**	**24.48**	**34.48**	**76.34**	**82.89**
Intermediate Urban Fringe	–	–	–	–	–	–	–	–	–	–	–	–
Barautha	73.37	68.99	31.33	43.98	7.08	12.34	1.67	9.19	19.55	18.67	67	46.84
Jarauth	92.96	76.7	21.99	54.10	3.52	8.81	0.83	7.38	3.52	14.49	77.18	38.52
Kothia	83.57	77.25	41.31	36.83	8.21	16.31	3.29	0.00	8.21	6.44	55.4	63.17
Nagaula	74.87	48.24	51.91	43.09	14.70	16.86	5.43	9.55	10.43	34.9	42.67	47.36
Bhankri Khas	80.58	86.85	45.82	33.53	3.56	1.81	2.44	4.19	15.86	11.34	51.73	62.28
Digsi	68.80	60.33	58.78	50.88	13.68	18.03	22.39	3.78	17.52	21.64	18.83	45.34
Imlani	70.72	76.65	63.25	27.87	17.76	13.71	2.14	0.88	11.51	9.64	34.62	71.25
Karsua	76.27	82.43	68.95	56.36	8.47	5.68	2.29	4.94	15.25	11.89	28.76	38.70
Kishan Garhi	75.94	56.35	53.08	62.97	8.02	22.1	1.37	1.27	16.04	21.55	45.55	35.76

(continued)

Table 3.3 (continued)

Sampled villages	Primary sector				Secondary sector				Tertiary sector			
	1981	1991	2001	2011	1981	1991	2001	2011	1981	1991	2001	2011
Jatpura	55.87	92.2	54.13	49.81	25.42	0	22.51	27.76	18.72	7.8	23.36	22.43
Alipur	80.71	91.87	66.2	49.07	8.04	4.07	10.42	4.63	11.25	4.07	23.38	46.30
Rafipur Siya	69.72	70.49	64.39	79.13	10.09	13.11	22.6	2.54	20.18	16.39	13.01	18.32
Badwamni	80.80	82.91	66.03	72.26	8.04	4.7	0.32	5.48	11.16	12.39	33.65	22.26
Bhartua	92.84	97.56	61.16	89.38	2.39	0.16	2.42	3.10	4.78	2.28	36.42	7.52
Madanpur Chhabila	95.00	89.74	68.23	65.75	0.71	4.49	1.56	2.74	4.29	5.77	30.21	31.51
Average	**80.19**	**75.7**	**54**	**52.96**	**9.70**	**7.62175**	**7.08**	**5.66**	**12.19**	**14.95**	**38.92**	**41.38**
Distant Urban Fringe	–	–	–	–	–	–	–	–	–	–	–	–
Maimari	82.00	81.49	74.25	38.17	2.28	4.36	5.8	7.70	15.72	14.16	19.95	54.13
Arrana	82.89	87.73	56.17	59.44	0.00	4.33	0.85	9.32	17.11	7.94	42.98	31.24
Chuharpur	60.11	57.2	45.87	54.33	16.94	12.29	11.01	1.73	22.95	30.51	43.12	43.94
Ahamadpuri	78.04	85.99	68.58	5.52	2.70	3.86	4.23	1.70	19.26	10.14	27.19	92.78
Tejpur	83.88	73.33	63.32	60.99	4.96	2.12	2.19	0.41	11.16	24.55	34.48	38.60
Sisroi Nagla Dhak	93.33	65.38	57.41	21.90	0.00	26.92	3.09	5.11	6.67	7.69	39.51	72.99
Peepali	96.20	83.25	61.19	61.36	0.00	7.18	3.36	6.82	3.80	9.57	35.45	31.82
Daud Garhi	74.05	55.09	47.3	42.69	11.45	30.54	12.84	1.75	14.50	14.37	39.86	55.56
Navalpur	97.79	86.05	92.77	95.06	0.00	1.74	0	0.00	2.21	12.21	7.23	4.94
Gada Khera	79.31	79.31	63.73	50.28	17.93	13.79	6.86	37.57	2.76	6.9	29.41	12.15
Sikargarhi	80.39	85.54	61.24	47.45	0.65	1.2	1.12	14.80	18.95	13.25	37.64	37.76
Jagatpur	99.13	99.44	63.04	90.09	0.00	0	22.83	0.00	0.87	0.56	14.13	9.91
Pratappur	93.53	76.15	75.11	92.80	0.00	3.21	1.29	2.31	6.47	20.64	23.61	4.88
Jangal Garhi	97.52	85.71	63.7	54.77	0.00	3.73	0.37	0.92	2.48	10.56	35.93	44.31
Padki	77.38	99.19	75.44	73.53	9.95	0.54	2.92	0.29	12.67	0.27	21.64	26.18

(continued)

Table 3.3 (continued)

Sampled villages	Primary sector				Secondary sector				Tertiary sector			
	1981	1991	2001	2011	1981	1991	2001	2011	1981	1991	2001	2011
Daupur	86.07	97.53	83.57	86.16	5.57	0	0	0.00	8.36	2.47	16.43	13.84
Bhataula	100.00	88.89	70.27	96.39	0.00	0.41	0	0.00	0.00	10.7	29.73	3.61
Parauri	88.90	62.22	69.27	43.84	0.13	5.11	0.23	4.89	10.96	32.67	30.5	51.27
Average	**84.94**	**80.7**	**66.01**	**53.69**	**3.58**	**5.4**	**4.02**	**4.99**	**11.48**	**13.91**	**29.97**	**41.32**
Average for whole UF	**76.87**	**66.11**	**46.02**	**28.2**	**7.48**	**12.78**	**5.57**	**4.69**	**15.64**	**21.11**	**48.41**	**67.11**

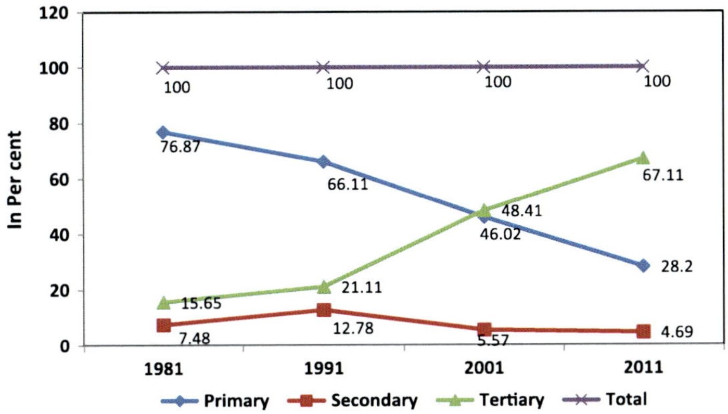

Fig. 3.1 Share of sector wise workers in urban fringe of Aligargh city. *Source* Population census of India, 1981, 1991, 2001 & 2011

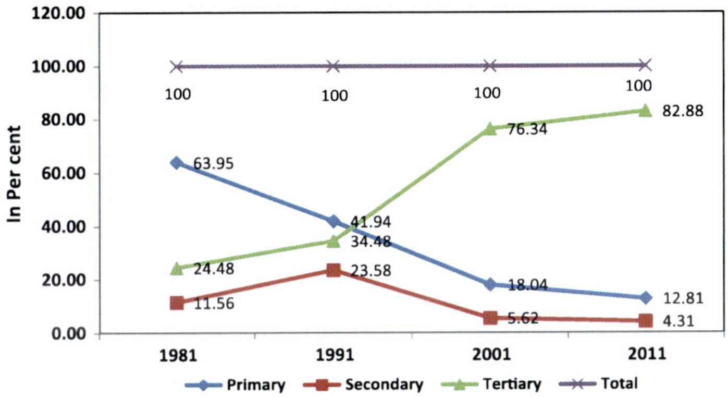

Fig. 3.2 Share of sector wise workers in immediate urban fringe. *Source* Population census of India, 1981, 1991, 2001 & 2011

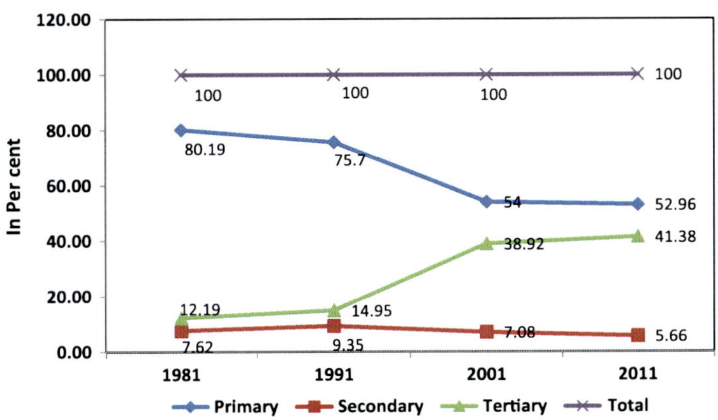

Fig. 3.3 Share of sector wise workers in intermediate urban fringe. *Source* Population census of India, 1981, 1991, 2001 & 2011

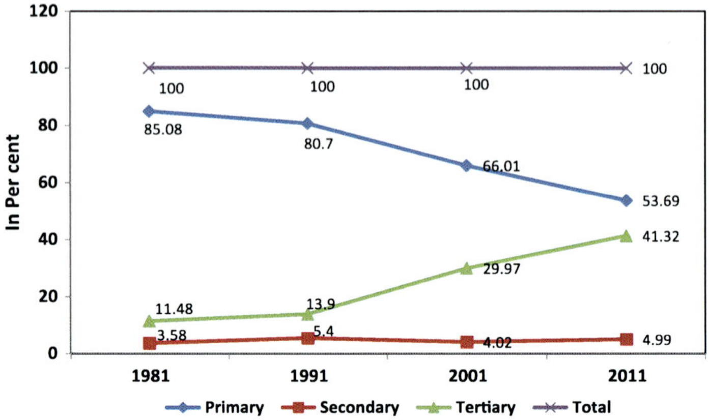

Fig. 3.4 Share of sector wise workers in distant urban fringe. *Source* Population census of India, 1981, 1991, 2001 & 2011

Diverse Livelihood Options in the Urban Fringe of Aligarh city

Repair shop

Cycle repair shop

Saloon shop

Tailoring shop

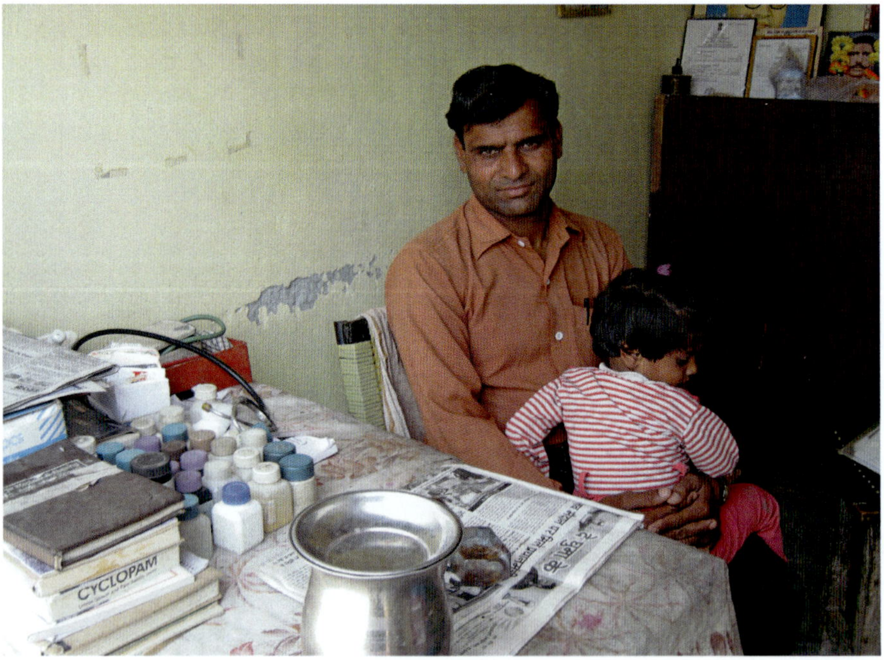

Quack

Urban labourers

3.4 Livelihood Diversification

A household may diversify its activities by increasing the number of workers in the household or by the participation of each member in more than one economic activity (Unni 1996). These both are the focus of the discussion here. Table 3.4 presents share of households engaged in different economic sectors. The survey reveals that households now generate income from more than one economic sector. The urban economy offers them opportunities of employment in secondary and tertiary sectors. This diversification in economic sectors was again found to be pronounced in immediate fringe (Figs. 3.5, 3.6 and 3.7).

A chi-square test for independence (Pearson Chi-Square value) indicated significant association between distance from the city centre and the number of economic sectors from which household's income generated, $\chi 2$ (1, n = 757) = 17.707, p = 0.000, Cramer's V = 0.15 (Appendix A). This indicates that the proportion of households from each urban fringe whose income generated from either one sector

Table 3.4 Aligarh urban fringe: household's association with different sectors economy

Zone		Number of sectors		Total
		Only one sector	More than one sector	
Immediate urban fringe	Count	**106**	**287**	**393**
	% within zone	27.0 %	73.0 %	100.0 %
	% within fringe	42.7 %	56.4 %	51.9 %
Intermediate urban fringe	Count	**67**	**131**	**198**
	% within zone	33.8 %	66.2 %	100.0 %
	% within fringe	27.0 %	25.7 %	26.2 %
Distant urban fringe	Count	**75**	**91**	**166**
	% within zone	45.2 %	54.8 %	100.0 %
	% within fringe	30.2 %	17.9 %	21.9 %
Total	Count	**248**	**509**	**757**
	% within zone	32.8 %	67.2 %	100.0 %
	% within fringe	100.0 %	100.0 %	100.0 %

Notes
Figures in bold are absolute number of sampled respondents in the particular distance zones
Figures in underlined are percentages to the total from that particular distance zone
Figures below the underlined figures are percentages to the total of that particular variable
Source Based on field survey (2013)

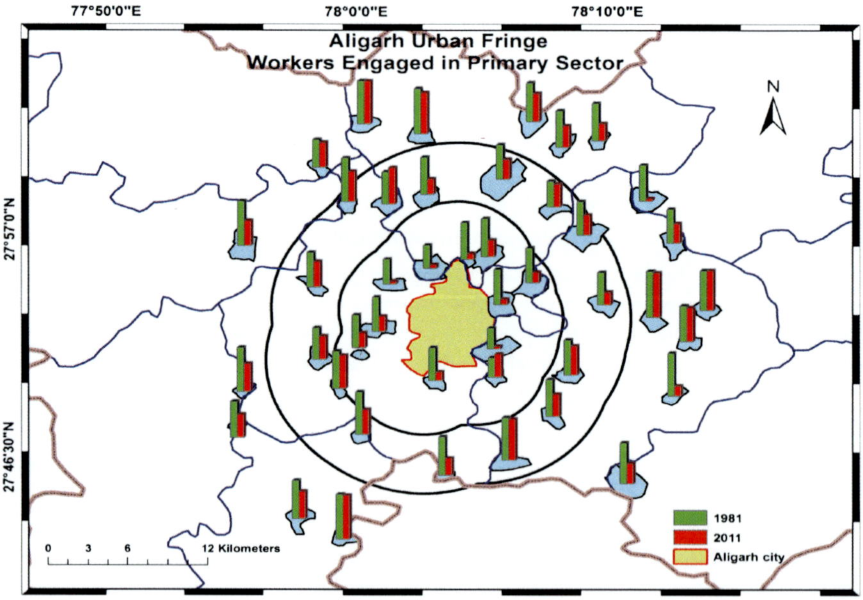

Fig. 3.5 Aligarh urban fringe: workers engaged in primary sector. *Source* Based on population census of India, 1981 & 2011

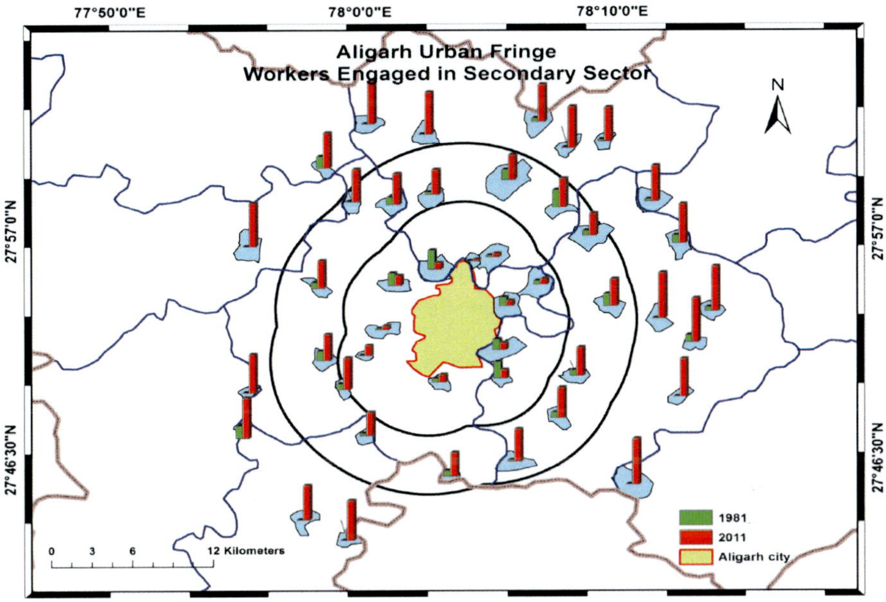

Fig. 3.6 Aligarh urban fringe: workers engaged in secondary sector. *Source* Based on population census of India, 1981 & 2011

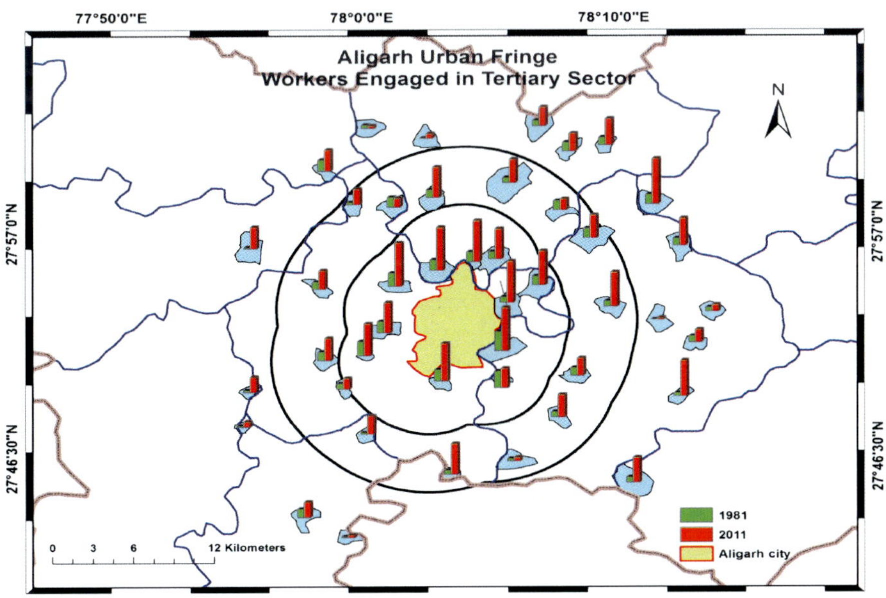

Fig. 3.7 Aligarh urban fringe: workers engaged in tertiary sector. *Source* Based on population census of India, 1981 & 2011

or more than one sector is significantly different from each other. The immediate urban fringe households get more option for diversifying their activities because of lesser distance.

The household's income from primary activities is higher in distant urban fringe. It is becoming less significant with city proximity. A great variety of activities draw labourers to work in secondary and tertiary sector in urban fringe. Dependency on only tertiary sector is significantly higher in immediate urban fringe because urban accessibility offers the opportunities. Even the household's income from both primary and tertiary is also greater in immediate urban fringe (Table 3.5). Employment in secondary sector is less important mainly because the opportunities are in unorganized sector.

A great variety of activities draw labourers to work in tertiary sector in the fringe areas. It is to be found that 69.1 % of the total sampled households engaged are associated with this sector (Table 3.5). In immediate urban fringe, 78.6 % households are associated with tertiary sector along with other sectors of economy and it is surprising that 21.2 % households are fully devoted to tertiary activities. The people are found to make more frequently adapt to tertiary activities in this zone than the intermediate and distant urban fringe. Most of the workers engaged in tertiary sector are employed in informal sector.

3.5 Economically Active Member Within Households

Participation of household's members in non-agricultural work cannot be adequately analyzed without undertaking relation between agricultural and non-agricultural activities. Moreover, the nature of participation of household members in economic activities is influenced by their access to land and other assets, etc. There is an emerging consensus that the livelihood security and well-being of rural households improve with the blending of non-farm economic activities with farm activities and such diversification of rural livelihood positively impacts the farm efficiency. The sample survey shows that nearly about two thirds of the total households reported of having more than one working member within household (Fig. 3.8). This suggests that more workers within household are engaged themselves in various type of economic activities to earn more and more to meet their growing needs and demands which are the results of city proximity and influence of urban way of lifestyle.

Table 3.5 Aligarh urban fringe: household's income generated from different sectors

Zone		Number of sector							Total
		P	S	T	P + S	P + T	S + T	P + S+T	
Immediate urban fringe	Count	**50**	**17**	**83**	**17**	**200**	**8**	**18**	**393**
	% within zone	12.7 %	4.3 %	21.1 %	4.3 %	50.9 %	2.0 %	4.6 %	100 %
	% within fringe	42.0 %	54.8 %	84.7 %	20.5 %	54.2 %	36.4 %	51.4 %	51.9 %
Intermediate urban fringe	Count	**26**	**9**	**11**	**22**	**115**	**8**	**7**	**198**
	% within zone	13.1 %	4.5 %	5.6 %	11.1 %	58.1 %	4.0 %	3.5 %	100 %
	% within fringe	21.8 %	29.0 %	11.2 %	26.5 %	31.2 %	36.4 %	20.0 %	26.2 %
Distant urban fringe	Count	**43**	**5**	**4**	**44**	**54**	**6**	**10**	**166**
	% within zone	25.9 %	3.0 %	2.4 %	26.5 %	32.5 %	3.6 %	6.0 %	100 %
	% within fringe	36.1 %	16.1 %	4.1 %	53.0 %	14.6 %	27.3 %	28.6 %	21.9 %
Total	Count	**119**	**31**	**98**	**83**	**369**	**22**	**35**	**757**
	% within zone	15.7 %	4.1 %	12.9 %	11.0 %	48.7 %	2.9 %	4.6 %	100 %
	% within fringe	100 %	100 %	100 %	100 %	100 %	100 %	100 %	100 %

Notes
i. Figures in bold are absolute number of sampled respondents in the particular distance zones
ii. Figures in underlined are percentages to the total from that particular distance zone
iii. Figures below the underlined figures are percentages to the total of that particular variable
iv. P is indicating primary sector of economy
v. S is indicating secondary sector of economy
vi. T is indicating tertiary sector of economy
Source Based on field survey (2013)

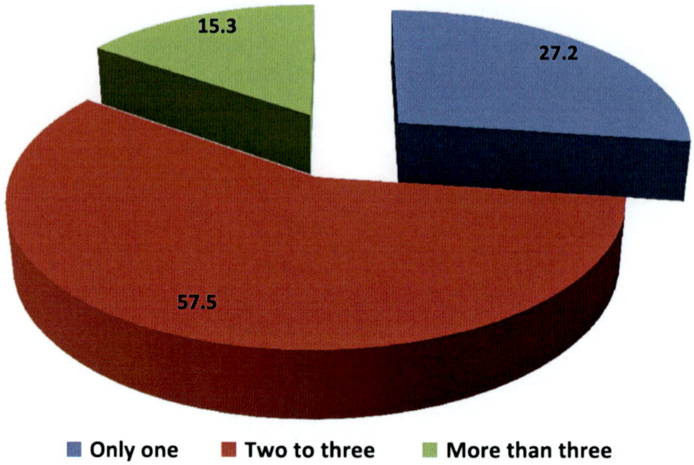

Fig. 3.8 Number of working members within a household. *Source* Based on field survey, (2013)

3.6 Age Preferred Economic Activities

The youth in urban fringe had some exposure to education. This is an important phenomenon of urban fringe, where education certainly has made inroads in the society and every section of the society has some level of education. But this education does not guarantee employment opportunities in urban area (especially white collar job, for which they aspire). This is a very critical situation among the urban fringe youth, leading to frustration, desperation and depression.

They have education (academic degrees) and they now do not want to associate themselves with traditional agricultural activity and working on farms. So, the easiest option is to open a shop with small capital. In most cases, their sustenance is difficult but they pursue with it to have a better social status, which is higher in trading than of soil cultivation. The Table 3.6 shows that out of the total sampled workers belong to 15–34 years age group, 67 % are engaged in secondary and tertiary activities. A chi-square test for independence (Pearson Chi-Square value) indicated significant association between proportion of workers in different sectors and their age, $\chi 2$ (1, $n = 1703$) = 84.20, $p = 0.000$, Cramer's V = 0.22. This means that the proportion of workers in primary sector whose age is significantly different from workers engaged in secondary and tertiary sectors. The Cramer's coefficient value is 0.22 indicating a highly strong association between the two variables (Appendix B).

Table 3.6 Aligarh urban fringe: sector wise working age group

Sector of economy		Age group			Total
		15-years	35–59 years	60 years above	
Primary sector	Count	**246**	**450**	**66**	**762**
	% within sector	32.3 %	59.1 %	8.7 %	100.0 %
	% within age group	32.6 %	55.7 %	46.8 %	44.7 %
Secondary and tertiary sector	Count	**508**	**358**	**75**	**941**
	% within sector	54.0 %	38.0 %	8.0 %	100.0 %
	% within age group	67.4 %	44.3 %	53.2 %	55.3 %
Total	Count	**754**	**808**	**141**	**1703**
	% within sector	44.3 %	47.4 %	8.3 %	100.0 %
	% within age group	100.0 %	100.0 %	100.0 %	100.0 %

Notes
i. Figures in bold are absolute number of sampled respondents in the particular sector of economy
ii. Figures in underlined are percentages to the total from that particular sector of economy
iii. Figures below the underlined figures are percentages to the total of that particular variable
Source Based on field survey (2013)

3.7 Generation-Wise Occupation

Urban dwellers engaged in urban agriculture, there are differences in terms of gender, age and educational status (Brook 2002). Studies have showed that younger population is involved in urban tertiary sector while older generation was found to be associated more with primary sector (ibid.).

The survey on the urban fringe of Aligarh city also shows clearly that there is an overall decline in agriculture in the families across the generations. Most of the first generation people or the elder population was reported to be engaged in farming activity. The survey reveals that in Aligarh's urban fringe, there is rapid decrease in second generation and only a few of third generation population are associated with agriculture (Fig. 3.9). This suggests that there is transformation of occupational choice, the elders of the household are engaged in farming but the younger generation are distancing themselves from farming activity and inclined towards urban oriented tertiary sector.

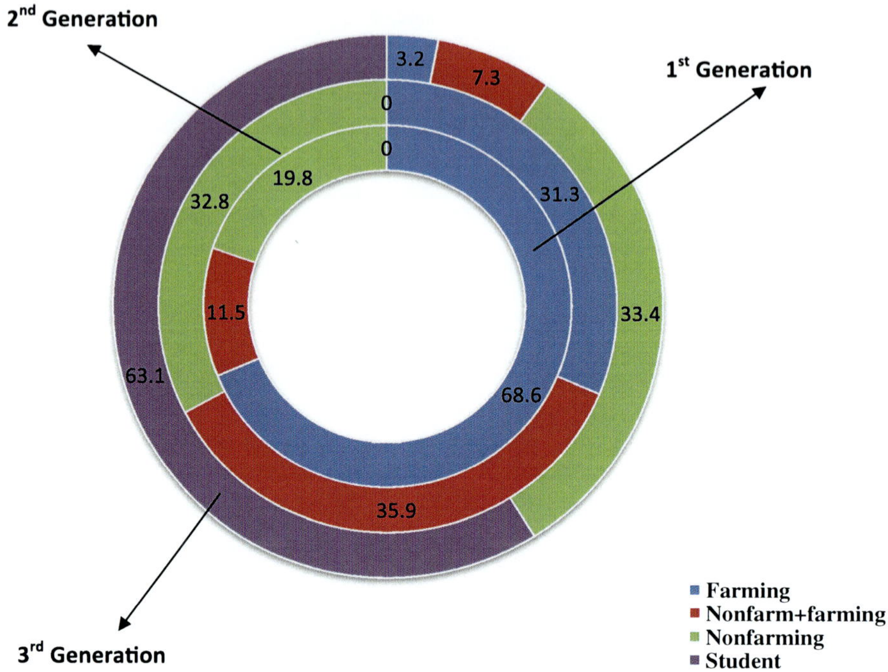

Fig. 3.9 Aligarh urban fringe: generation-wise occupation among sampled households. *Source* Based on field survey, (2013)

3.8 Status of Women

Women's work participation in Indian context has poor representation generally. There are social and cultural limitations for women employment. Traditionally women employment outside their house is not acceptable. However, with the process of modernization and urbanization, this tradition is disappearing, especially for off-farm activity. But the domestic duties of women also restrict them to work outside their homes. This is clearly visible from the Table 3.7 that over the last three decades the share of female workers to total work force has not changed much. The census data (Table 3.8) shows that in 2011, three fourth workers out of the total working female in the sampled villages were engaged in secondary and tertiary sector but still 23 % were engaged in on-farm activities. Women from immediate urban fringe are found more to be work outside agricultural sector (Table 3.9).

Table 3.7 Aligarh urban fringe: share of women workers to total main working population

Zone	1981			1991			2001			2011		
	FW[a]	TW[b]	FW (%)	FW[a]	TW[b]	FW (%)	FW[a]	TW[b]	FW (%)	FW[a]	TW[b]	FW (%)
Immediate urban fringe	92	3684	2.5	335	7065	4.74	821	11372	7.22	348	20690	1.68
Intermediate urban fringe	87	4579	1.9	126	5430	2.32	743	5878	12.64	919	6416	14.32
Distant urban fringe	139	4017	3.46	65	4745	1.37	554	4307	12.86	1257	6254	20.10
Total	**318**	**12280**	**2.59**	**826**	**17340**	**4.76**	**2118**	**21557**	**9.83**	**4579**	**33360**	**13.73**

Source Based on Population Census of India, 1981, 1991, 2001 & 2011

[a]FW = Female Workers

[b]TW = Total Workers

Table 3.8 Aligarh urban fringe: women work participation in the sampled villages

Zone	Primary sector				Secondary sector				Tertiary sector			
	1981	1991	2001	2011	1981	1991	2001	2011	1981	1991	2001	2011
Immediate urban fringe	**36**	**210**	**254**	**348**	**5**	**28**	**140**	**165**	**51**	**97**	**427**	**1890**
	39.13	62.69	30.94	14.48	5.43	8.36	17.05	6.87	55.43	28.96	52.01	78.65
	19.15	33.49	34.46	32.55	38.46	87.50	34.91	37.33	43.59	58.08	43.57	61.60
Intermediate urban fringe	**50**	**72**	**257**	**320**	**6**	**4**	**200**	**93**	**31**	**50**	**286**	**506**
	57.47	57.14	34.59	34.82	6.90	3.17	26.92	10.12	35.63	39.68	38.49	55.06
	26.60	11.48	34.87	29.93	46.15	12.50	49.88	21.04	26.50	29.94	29.18	16.49
Distant urban fringe	**102**	**345**	**226**	**401**	**2**	**0**	**61**	**184**	**35**	**20**	**267**	**672**
	73.38	94.52	40.79	31.90	1.44	0.00	11.01	14.64	25.18	5.48	48.19	53.46
	54.26	55.02	30.66	37.93	15.38	0.00	15.21	41.63	29.91	11.98	27.24	21.90
Total	**188**	**627**	**737**	**1069**	**13**	**32**	**401**	**442**	**117**	**167**	**980**	**3068**
	59.12	75.91	34.80	23.35	4.09	3.87	18.93	9.65	36.79	20.22	46.27	67.00
	100	100	100	100	100	100	100	100	100	100	100	100

Notes
i. Figures in bold are absolute number of sampled respondents in the particular distance zones
ii. Figures in underlined are percentages to the total from that particular distance zone
iii. Figures below the underlined figures are percentages to the total of that particular variable
Source Based on Population Census of India, 1981, 1991, 2002 & 2011

Table 3.9 Aligarh urban fringe: working population engaged in different economic activities

Zone		Type of occupation of female workers in the urban fringe of Aligarh city										Total
		Cultivation	Agricultural labour	Dairying	Construction labour	HH industry	Non-HH industry	Trade & business	Service	Trans. & comm.	Other	
Immediate urban fringe	Count	10	0	111	15	10	37	0	22	2	0	207
	% within zone	4.8 %	0.0 %	53.6 %	7.2 %	4.8 %	17.9 %	0.0 %	10.6 %	1.0 %	0.0 %	100.0 %
	% within fringe	10.6 %	0.0 %	56.9 %	57.7 %	100.0 %	100.0 %	0.0 %	56.4 %	100.0 %	0.0 %	47.6 %
Intermediate urban fringe	Count	41	5	54	7	0	0	0	8	0	3	118
	% within zone	34.7 %	4.2 %	45.8 %	5.9 %	0.0 %	0.0 %	0.0 %	6.8 %	0.0 %	2.5 %	100.0 %
	% within fringe	43.6 %	21.7 %	27.7 %	26.9 %	0.0 %	0.0 %	0.0 %	20.5 %	0.0 %	42.9 %	27.1 %
Distant urban fringe	Count	43	18	30	4	0	0	2	9	0	4	110
	% within zone	39.1 %	16.4 %	27.3 %	3.6 %	0.0 %	0.0 %	1.8 %	8.2 %	0.0 %	3.6 %	100.0 %
	% within fringe	45.7 %	78.3 %	15.4 %	15.4 %	0.0 %	0.0 %	100.0 %	23.1 %	0.0 %	57.1 %	25.3
Total	Count	94	23	195	26	10	37	2	39	2	7	435
	% within zone	21.6 %	5.3 %	44.8 %	6.0 %	2.3 %	8.5 %	0.5 %	9.0 %	0.5 %	1.6 %	100.0 %
	% within fringe	100.0 %	100.0 %	100.0 %	100.0 %	100.0 %	100.0 %	100.0 %	100.0 %	100.0 %	100.0 %	100.0

Notes
i. Figures in bold are absolute number of sampled respondents in the particular distance zones
ii. Figures in underlined are percentages to the total from that particular distance zone
iii. Figures below the underlined figures are percentages to the total of that particular variable
Source Based on field survey (2013)

Aligarh Urban Fringe: Diverse Livelihood Options for Women

Women are working as agricultural labourers

Women contribute significantly to their household income by performing side jobs for their male counterparts

Engaged in handicrafts work

Engaged in household industry

Woman has greater role as home maker

Education has helped women in getting jobs in tertiary sector

References

Adu Ampong EF et al. (2008) Socio-economic transitions: changing livelihoods in the peri-urban interface, a case study in peri-urban interface of Kumasi. Draft Report (July 25, 2008), Tropenbos International, Ghana, Amsterdam University, College of Agriculture and Natural Resource, KNUST

Brook RM (2002) Natural resource management action plan development for Hubli Dharwad peri urban interface. Final Technical Report for project R7959, School of Agricultural and Forest Sciences, University of Wales, Bangor, UK

Ministry of Urban Development, Government of India (2009) City sanitation plan—Aligarh. www.urbanindia.nic.in/programme/uwss/CSP/Draft.../Aligarh_CSP.pdf

Unni J (1996) Diversification of economic activities and non-agricultural employment in rural Gujarat. Econmic Polit Wkly 31:33

Chapter 4
Urban Fringe: Crop Cultivation as Livelihood Option

The urban fringe, surrounding urban areas is one of the most critical areas of human acumen. It is not the extent of the resource base contained in the urban fringe that makes it critical to society, but rather the fact that the land and resources therein are subject to competing, often conflicting, demands. The urban fringe has been placed under pressure increasingly with increased demand due to rapidly changing social, cultural and economic values associated with land and its resources. The rapid sprawl of cities, outside its political boundaries and beyond its real physical—cultural domain possesses several attendant conflicts and stresses and has also stimulated various adaptations of human activities. These are reflected through diverse facets, especially changing urban fringe and adjacent or interstitial rural land uses.

Agriculture is one of the primitive economic activity, but with cultural development of mankind, the dependency for livelihood attainments have shifted to secondary and tertiary sectors. Agriculture is still an important sector of the Indian economy as it contributes about 25 % of the total Gross Domestic Product (GDP) and provides employment to about 67 % of the working population of the country. Over the past two decades, the service sector has replaced agriculture as the main growth sector in India, thrusting the focus of development into the city (Chanda 2002). But in the transforming economic environment, agricultural activity is no longer a preferred livelihood option as the census data also indicates that agriculture workers decreased by 20 % in the last three decades. Economic transformations offered opportunities for local economic development in peripheral areas of urban bodies (Keivani and Mattingly 2007). In many cases, such processes lead to an expansion of formal physical development activity on peripheral land, and/or the restructuring of existing development there (Mehta 2009).

In this chapter, the focus has been given on crop cultivation as livelihood and an attempt was made to analyze the cropping pattern, factors associated with this activity and emerging trends and patterns of farming activity in Aligarh's urban fringe.

N. Banu and S. Fazal, *Livelihood and Wellbeing in the Urban Fringe*,
The Urban Book Series, DOI 10.1007/978-3-319-39660-6_4

4.1 Crop Cultivation as Livelihoods (At Individual Level)

The census data shows that out of the total workers only 13 % is associated with agriculture in immediate urban fringe and this share is increasing with distance from the city as in distant fringe still 54 % workers are practicing agriculture (Table 4.1). Table 4.2 shows that the decadal growth rate is negative for farming activities in whole urban fringe and highest negative growth is to be found throughout the last three decades in immediate urban fringe. As the urban influence is greater on immediate urban fringe than the distanced zones, the negative growth rate in agricultural activities is increasing with city proximity (Table 4.2).

Table 4.1 Aligarh urban fringe: share of agricultural worker to main worker in sampled villages

Zone	Agricultural worker				Total			
	1981	1991	2001	2011	1981	1991	2001	2011
Immediate urban fringe	**2356**	**2963**	**2052**	**2650**	**3684**	**7065**	**11372**	**20690**
	63.95	41.94	18.04	12.81	–	–	–	–
Intermediate urban fringe	**3672**	**4186**	**3174**	**3398**	**4579**	**5430**	**5878**	**6416**
	80.19	75.70	54.00	52.96	–	–	–	–
Distant urban fringe	**3412**	**3830**	**2753**	**3358**	**4017**	**4745**	**4307**	**6254**
	84.94	80.72	63.92	53.69	–	–	–	–
Total	**9440**	**10979**	**7979**	**9406**	**12280**	**17340**	**21557**	**33360**
	76.87	63.31	37.01	28.20	–	–	–	–

Notes
i. Figures in bold are absolute number of sampled respondents in the particular distance zones
ii. Figures in underlined are percentages to the total from that particular distance zone
Source Census of India (1981, 1991, 2001 & 2001)

Table 4.2 Aligarh urban fringe: growth of agricultural workers (1981–2011)

Zone	1981	1991	2001	2011	1981–91	1991–01	2001–11
	Agri.	Agri.	Agri.	Agri.	Growth rate	Growth rate	Growth rate
Immediate urban fringe	63.95	41.94	18.04	12.81	−34.42	−56.99	−29.00
Intermediate urban fringe	80.19	75.7	54.00	52.96	−5.60	−28.67	−1.92
Distant urban fringe	84.94	80.7	66.01	53.69	−4.99	−18.20	−18.65
Total	76.87	63.31	37.01	28.20	−17.64	−41.54	−23.81

Source Primary Census Abstract, 1981, 1991, 2001 & 2011

4.2 Crop Cultivation as Livelihoods (At Household Level)

The sample survey in Aligarh's urban fringe was spread over a total of 44 villages. A total of 757 households were sampled out of which 457 (60 %) households were indulged in primary activities (Table 4.3). This suggests that primary activity in Aligarh's urban fringe is still the main livelihood component among sampled households. Significantly, only 76 (10.0 %) households depended on crop cultivation alone, while remaining 381 (50.0 %) were supplementing with dairying activity along with crop cultivation. This shows that households in Aligarh's urban fringe were diversifying their livelihoods to enhance their earnings. This argument is also strengthened by the fact that this trend was found to be more dominant among households in villages closer to the city. The survey shows that 160 (40.7 %) households in *immediate* urban fringe were engaged in both dairying and as well as crop cultivation activity while only 21 (5.3 %) of households in this zone depended only on crop cultivation. However, the crop cultivation becomes more

Table 4.3 Aligarh urban fringe: household engagement in different economic activities

Zone		Type of economic activity				Total
		Only cultivation	Only dairy	Cultivation & dairy	Non agriculture	
Immediate urban fringe	Count	**21**	**40**	**160**	**172**	**393**
	% within zone	5.3 %	10.2 %	40.7 %	43.8 %	100.0 %
	% within fringe	27.6 %	64.5 %	42.0 %	72.9 %	52.1 %
Intermediate urban fringe	Count	**29**	**14**	**107**	**48**	**198**
	% within zone	14.6 %	7.1 %	54.0 %	24.3 %	100.0 %
	% within fringe	38.2 %	22.6 %	28.1 %	19.5 %	26.0 %
Distant urban fringe	Count	**26**	**8**	**114**	**18**	**166**
	% within zone	15.7 %	4.8 %	68.7 %	10.8 %	100.0 %
	% within fringe	34.2 %	12.9 %	29.9 %	7.6 %	22.0 %
Total	Count	**76**	**62**	**381**	**238**	**757**
	% within zone	10.0 %	8.2 %	50.0 %	31.4 %	100.0 %
	% within fringe	100.0 %	100.0 %	100.0 %	100.0 %	100.0 %

Notes
i. Figures in bold are absolute number of sampled respondents in the particular distance zones
ii. Figures in underlined are percentages to the total from that particular distance zone
iii. Figures below the underlined figures are percentages to the total of that particular variable
Source Based on field survey (2013)

prominent with the distance from the city as *intermediate* urban fringe had 15 % and *distant* urban fringe reported 16 % households associated with only crop cultivation.

4.3 Nature of Farming Activity

The sample survey of 457 households associated with farming activity reveals that there was fair degree of commercialization in their economic activities. It was found that almost half of these 457 households were first fulfilling their own requirements and surplus primary productions were disposed off in the market to get cash returns. Another 21 % were producing commodities only for markets. These producers were selling their entire productions to market for cash returns. The remaining 30 % were subsistence producers; they have reported to consume most of their productions themselves and very rarely they are left with surplus productions for market disposal (Table 4.4). These subsistence producers are mostly small land owners,

Table 4.4 Aligarh urban fringe: type of cropping system

Zone		Type of cropping system			Total
		Purely subsistence	Subsistence and commercial	Purely commercial	
Immediate urban fringe	Count	**14**	**105**	**62**	**181**
	% within zone	7.7 %	58.0 %	34.3 %	100.0 %
	% within fringe	10.1 %	47.7 %	63.3 %	39.6 %
Intermediate urban fringe	Count	**44**	**61**	**31**	**136**
	% within zone	32.4 %	44.9 %	22.8 %	100.0 %
	% within fringe	31.7 %	27.7 %	31.6 %	29.8 %
Distant urban fringe	Count	**81**	**54**	**5**	**140**
	% within zone	57.9 %	38.6 %	3.6 %	100.0 %
	% within fringe	58.3 %	24.5 %	5.1 %	30.6 %
Total	Count	**139**	**220**	**98**	**457**
	% within zone	30.4 %	48.1 %	21.4 %	100.0 %
	% within fringe	100.0 %	100.0 %	100.0 %	100.0 %

Notes
i. Figures in bold are absolute number of sampled respondents in the particular distance zones
ii. Figures in underlined are percentages to the total from that particular distance zone
iii. Figures below the underlined figures are percentages to the total of that particular variable
Source Based on field survey (2013)

where they are unable to produce enough to meet with their requirements. Their economic activity has subsistence orientation largely due to small (uneconomic) land holdings which restricts them in commercializing. Significantly, distance from city has crucial bearing on nature of farming activity. The sample survey suggests that closer to the city have higher degree of commercialization and decreases as we move away from city. There were negligible subsistence producers in the *immediate* urban fringe; on the contrary 58 % of the subsistence producers were in *distant* urban fringe. This suggests that closer to the city, there is intense urban forces resulting in higher degree of commercialization.

4.4 Farm Size

A total of 457 households out of 757 sampled households have reported the association with farming activity. Out of these 457 households, 56 were categorized as large farmers (farm size of more than 10 ha), 138 were categorized as medium farmers (farm size of 4–10 ha) and remaining 263 were categorized as small farmers (farm size of less than 4 ha) (Table 4.5). Spatially, the sample survey finds that in the *immediate* urban fringe (closer to the city) the land holdings are smaller

Table 4.5 Aligarh urban fringe: farm size among sampled households

Zone		Farm size			Total
		Large	Medium	Small	
Immediate urban fringe	Count	**4**	**43**	**134**	**181**
	% within zone	2.21 %	23.76 %	74.03 %	100.00 %
	% within fringe	7.14 %	31.16 %	50.95 %	39.60 %
Intermediate urban fringe	Count	**19**	**49**	**68**	**136**
	% within zone	13.97 %	36.03 %	50.00 %	100.00 %
	% within fringe	33.92 %	35.50 %	25.85 %	29.80 %
Distant urban fringe	Count	**33**	**46**	**61**	**140**
	% within zone	23.57 %	32.85 %	43.57 %	100.00 %
	% within fringe	58.92 %	33.33 %	23.19 %	30.63 %
Total	Count	**56**	**138**	**263**	**457**
	% within zone	12.25 %	30.20 %	57.54 %	100.00 %
	% within fringe	100.00 %	100.00 %	100.00 %	100.00 %

Notes
i. Figures in bold are absolute number of sampled respondents in the particular distance zones
ii. Figures in underlined are percentages to the total from that particular distance zone
iii. Figures below the underlined figures are percentages to the total of that particular variable
Source Based on field survey (2013)

and with increase in distance from the city, the size of land holdings are bigger. This suggests that closer to the city, there is intense pressure on land by urban forces to accommodate expanding urban population, resulting in fragmentation of land.

4.5 Marketing of Produced Crops

Another important information which was collected from the Aligarh's urban fringe was about modifications in farming activity. It was obvious that proximity to city would have greater urban and market forces as well as higher level of commercialization. The farmers would also try to maximize their returns and explore newer options for drawing maximum returns. The sample survey suggests that 27.6 % respondents were concern for better marketing and management for their farm products. Especially, producers from *immediate* urban fringe acting in urban system were found to be exploring most for possibilities of better marketing and management for their products. The urban fringe producers also showed interests in extension services offered by governmental agencies for upgrading their traditional skills.

Producers closer to the city can directly market their products by-passing the middlemen, thus obtaining even higher revenues. Households associated with perishable crops like vegetables and flower were found to be marketing their products only in the city market where its demand is more. Those households who market their perishable crops directly in the market, out of them 86 % households were from immediate urban fringe (Table 4.6). Disposal in mandi is becoming common with increasing distance from the city and in mandi the involvement of middlemen as local traders and city whole sellers was high. The sample survey shows that 28 % households in the urban fringe disposed their crops to the local traders in their nearby mandi and it was more pronounced in distant urban fringe where in 37 % households had to dispose it to the middlemen in the mandis (Table 4.6).

Thus with increasing distance from the city the share of profits is decreasing. Again a chi-square test for independence (Pearson Chi-Square value) indicated significant association between distance from the city center and the type of disposal of produced crops in the fringe, $\chi 2$ $(1, n = 457) = 186.45$, $p = 0.000$, Cramer's V = 0.45. The immediate urban fringe cultivators have better access for disposal of crops because lesser distance from the city makes it easier for immediate urban fringe producers to visit the city personally to dispose their products (Appendix C)

Table 4.6 Aligarh urban fringe: disposal of produced crop from sampled households

Zone		Disposal of Crop					Total
		Mandi	Mandi & city	Only city	From own farm	Do not dispose	
Immediate urban fringe	Count	**32**	**60**	**67**	**8**	**14**	**181**
	% within zone	17.7 %	33.1 %	37.0 %	4.4 %	7.7 %	100.0 %
	% within fringe	25.0 %	58.8 %	85.9 %	80.0 %	10.1 %	39.6 %
Intermediate urban fringe	Count	**44**	**35**	**11**	**2**	**44**	**136**
	% within zone	32.4 %	25.7 %	8.1 %	1.5 %	32.4 %	100.0 %
	% within fringe	34.4 %	34.3 %	14.1 %	20.0 %	31.7 %	29.8 %
Distant urban fringe	Count	**52**	**7**	**0**	**0**	**81**	**140**
	% within zone	37.1 %	5.0 %	0.0 %	0.0 %	57.9 %	100.0 %
	% within fringe	40.6 %	6.9 %	0.0 %	0.0 %	58.3 %	30.6 %
Total	Count	**128**	**102**	**78**	**10**	**139**	**457**
	% within zone	28.0 %	22.3 %	17.1 %	2.2 %	30.4 %	100.0 %
	% within fringe	100.0 %	100.0 %	100.0 %	100.0 %	100.0 %	100.0 %

Notes
i. Figures in bold are absolute number of sampled respondents in the particular distance zones
ii. Figures in underlined are percentages to the total from that particular distance zone
iii. Figures below the underlined figures are percentages to the total of that particular variable
Source Based on field survey (2013)

4.6 Leasing of Agricultural Land

There is intense pressure of urban and market forces on farming activity, which over the period of time has become stressed livelihood. The survey shows that 17.07 % households lease their land to their relatives or other farmers in terms of money for a particular period and do not engage themselves directly with cultivation of their land (Table 4.7). The land leasing activity shows that urban fringe households have lack of interest for farming activity although they own land. The survey shows that 26.51 % households in immediate urban fringe leasing their land and the trend of giving land on lease was increasing with city proximity because they were forced to dispose their agricultural land in their resident villages due to city's expansion on this zone and 40 % of those households who disposed their

Table 4.7 Aligarh urban fringe: land lease practice among the sampled households

Zone		Land lease		Total
		Taking land on lease	Giving land on lease	
Immediate urban fringe	Count	**15**	**48**	**181**
	% within zone	8.28 %	26.51 %	100.0 %
	% within fringe	22.38 %	61.54 %	39.60
Intermediate urban fringe	Count	**23**	**18**	**136**
	% within zone	16.91 %	13.24 %	100.0 %
	% within fringe	34.32 %	23.08 %	29.76
Distant urban fringe	Count	**29**	**12**	**140**
	% within zone	20.71 %	8.57 %	100.0 %
	% within fringe	43.28 %	15.38 %	30.63
Total	Count	**67**	**78**	**457**
	% within zone	14.66 %	17.07 %	100.0 %
	% within fringe	100.0 %	100.0 %	100.0 %

Notes
i. Figures in bold are absolute number of sampled respondents in the particular distance zones
ii. Figures in underlined are percentages to the total from that particular distance zone
iii. Figures below the underlined figures are percentages to the total of that particular variable
Source Based on field survey (2013)

land, again have bought agricultural land in distanced villages and it was difficult task for them to cultivate it by own self. On the other hand, the trend of taking land on lease was decreasing with city proximity and only 8.28 % sampled households were taking land on lease for cultivation, while it was 20.71 % in distant urban fringe. This is because more city proximity offers more opportunities for nonfarm activities.

4.7 Farm Activity as Part-Time Livelihood Option

The survey also reveals that even if the household is associated with farming activity, they often combine farming activities with other urban occupations, generating off farm income. Agricultural production in urban areas is rarely the only livelihood activity of a household (Prain 2006). The survey reveals that household's association with agriculture was quietly different from that of rural areas. Out of the total sample surveyed, 82.93 % cultivated their lands by own self with assistance of other agricultural labourers. The landless labours do not have to depend upon the agriculture completely and this group has been wiped out from urban fringe because of opportunities of economic activities in dairying activity and nonfarm sectors especially in city[1]. The overall effect of the change is that people have additional employment opportunities during nonharvest seasons in agriculture, and have good access to transportation facilities. In Aligarh's urban fringe, it was also found that the male members are engaged in farming activities in the village during peak season especially sowing and harvesting period and rest of the time they are depended on urban-based economic activities which provide higher and assured wages than the wages in rural agricultural sector. This phenomena is not only associated with landless labour but also with the small farmers. The small land holdings used for agricultural purpose make the farmers unable to sustain their livelihoods. This factor pushes the small farmers to engage themselves in non-farming activities to meet their needs and demands.

4.8 Intensification of Agriculture

Studies have shown that the farmers in the urban fringe areas have shifted from extensive to intensive agriculture (Coche 1982 and Mandere et al. 2010). Agriculture is typically becoming more intensive, with increasing emphasis on production of high value and perishable products such as vegetables, milk, eggs and fish, each with a ready urban market (Brook et al. 2000). This study also reveals that with the increasing intensity of commercialization, the agriculture has become intensive in the urban fringe. The study area lies in a relatively prosperous region of the country, where inputs of green revolution were adopted successfully. That is why the farms of Aligarh's urban fringe have reported higher level of application for agricultural inputs. The basic inputs of green revolution like irrigation (94.7 %), HYV seeds (87.1 %) chemical fertilizers (82.7 %) etc., were widely used by the sampled cultivators. But again, the application of these inputs were found to be more in farms closer to city, while its application declines with distance (Table 4.8). Apart from successful implementation of green revolution, other factors such as greater demand from markets combined with improved marketing channels, increased competition for land and access to production inputs (such as urban

[1]Per day 10,000 workers come from UF to city at the confluence of roads near Aligarh Development Authority Office and also at Dodhpur Chowraha. From here they are hired by builders or city people for constructional and other activities.

wastes and wastewater) and to credit are other important factors stimulating intensification in urban production. The inputs are quite expensive and apart from economically well off farmers, it is very difficult for most of the farmers to invest heavily for their crops. Immediate urban fringe farmers are more advantageous with easy access to financial institutions due to city proximity. Especially, the use of insecticides and pesticides were found to be used more in immediate urban fringe. This is mainly because of production of vegetables and flowers in this zone, similarly, the use of mechanical equipments (tractors, harvesters, cutter[2], etc.) have been reported a little more in distant urban fringe mainly because of relatively larger landholdings and greater volume of cereal production.

Aligarh Urban Fringe: The agricultural activity has been transformed from subsistence to commercial farming with inputs of green revolution being widely used

Use of pesticides have increased significantly for cash crops

[2]Cutter is modified form of thrashers.

Human labour has been replaced by mechanical equipments- ploughing by tractor

Irrigation through pump set

4.9 Cropping Pattern

The study area lies in the fertile tract of Ganga–Yamuna Doab region of India which has suitable condition for growing variety of crops. The region also has relatively better infrastructural and marketing facilities. All of these help in better farming activities in the study region. But farming in urban fringe is also influenced by urban forces evolving a different kind of cropping pattern and some crops like vegetables, peas, *arhar,* flowers (roses and marigold), *chara,* etc. have substantial acreage; these commodities are basically grown to meet the urban demands. Whereas, farms at *distant* urban fringe have shown subsistence orientation in their cropping pattern, as the cropping pattern was dominated by food grains.

The sample survey of 457 households involved in farming activity reported to own a total of 42335 ha of land. The entire land is used for agriculture in *rabi* season while in *kharif* season 94 % (39794 ha) of total land was under agriculture, leaving the remaining land as fallow. There are two main agricultural seasons such as *rabi* and *kharif.* Rabi season pertains to the winter period; it is the main dry crop season, while *kharif* season pertains to the rainy period. The main crops in rabi season are wheat, mustard, peas and gram, while in kharif season the important crops are rice, maize and arhar. Different varieties of vegetables, flower, fruits and chara are reported to be grown all round the year.

Agricultural census of India confirmed the aforesaid revealed from primary data. It revealed that total operational landholdings under agriculture decreased in urban fringe whereas it increased to 4.2 % in rest part of the district. This suggests the fact that there is intense pressure on agricultural land in the fringe area. But the land under cereals, vegetables and flowers increased. The survey reveals that cereal occupies maximum share of land followed by pulses and oilseeds. This is because, cereals have greater demands especially for the two principal crops, i.e. rice and wheat. These were also treated as favoured crop by the urban fringe farmers and market demands make these as commercial crops. Substantial area is also devoted to vegetables, flower and chara (flower and chara were categorized as other crops). These are prominent in immediate urban fringe as these crops are perishable as well as they have greater demand in the city.

4.9.1 Features of Farming Activity in Immediate Urban Fringe

The cropping pattern in the immediate urban fringe is dominated by cereals (34 %) followed by oilseeds (15 %), pulses (10 %), but significantly 28 % of the total land were devoted to vegetables and 13 % to other cash crops (flowers, chara, etc.)

(Table 4.9). It was noticed that farmers in immediate UF grow largely cereals for their subsistence requirements. Here they are growing cereals which require lesser inputs, labour and investments (wheat and maize needs lesser inputs than rice). Farmers in immediate UF faces the urban shadow effect the most, which in all sense affects agricultural activity in immediate UF. But despite urban forces, the farmers in immediate UF were also benefited from urban proximity. The cropping intensity, efficiency and provision of inputs were relatively higher in Aligarh's UF. The farmers in immediate UF were found to be most responsive to factors such as agricultural commodity price, yield and crop prospects. These factors reflect the high degree of commercialization in immediate UF farms. The main problem for farming activity in immediate UF was reported to be nonavailability of timely and adequate labourers. The farmers in immediate UF manage to get funds relatively easily and that is why availability of inputs is not a big problem for them. They also take risks and grow crops like vegetables, flowers, chara, etc. which they grow for the urban market. These crops are very important for farmers in immediate UF; their survival is greatly dependent on these crops, which also fetch them good income on daily basis. The striking feature of vegetable growers in immediate UF is that it depends on urban sewage or waste water for irrigating these farms (there are five drains around the city, which carries city sewage and waste water).

4.9.2 Features of Farming Activity in Intermediate Urban Fringe

Farmers in intermediate UF were reported to devote 52 % of their land to the cultivation of cereals followed by vegetables (15.6 %) and oilseeds (14.4 %). The farmers in intermediate UF also have good share of their land under cash crops because of their remunerative nature and assured urban market (Table 4.9). The share of urban market oriented crops among farmers in intermediate UF is also high, although these crops need high inputs and the chances of crop failure is always there but these farmers take risks and in return they get comparatively higher profits. Vegetables have substantial share in their cropping pattern in both rabi and kharif season, even other crops such as peas, gram and arhar can be categorised as vegetable as they have same demand and use in the city. Another crop is chara grown in kharif season and its acreage is also influenced by urban proximity. Chara is basically a fodder crop, which has significant demand as the region has well developed household dairy activity and farmers supplement their income by domesticating some cattle (buffaloes and cows). Almost 80 % of the chara grown by the sampled farmers are grown on large farms. Farmers in intermediate UF are the most progressive farmers of the Aligarh's UF. These farmers take agriculture activity seriously and consider it as a business and their livelihood is largely based on agriculture. They take advantage of relative proximity to the city as all the adverse impact of urbanization process on agriculture is absorbed by the farmers in immediate UF. Thus the farmers in this UF zone are at advantage both in terms of site and situation. They provide all the necessary inputs, that is why the yield in their farms are higher than others. The farmers of intermediate UF were found to be most responsive to factors such as agricultural commodity price, yield and crop prospects. These factors reflect the high degree of commercialization in intermediate UF farms. The main problem for farming activity in intermediate UF was reported to be nonavailability of timely and adequate provisions of irrigation, HYV seeds and labourers. These farmers were also concerned of government policies and plans and make their farming decisions keeping an eye on governmental policies.

4.9.3 Features of Farming Activity in Distant Urban Fringe

The farmers in distant UF were found to be allocating over two third of their total land to cereals (67.9 %), which are used as the staple food for the family but these cereal crops are also remunerative to grow because they have assured market and government and semi-government agencies procure cereals from their farms (Table 4.9). Rest of the crops grown by farmers in distant UF clearly indicates lesser influence of urban markets.Apart from cereals (wheat, rice and maize), the other crops are gram, arhar and mustard which are although commercial crops but they are not grown for direct disposal for city requirements. Vegetables are also grown but for self-consumption only and not for urban market. It was observed that farmers from distant UF were affected by the cost, adequate and timely availability of irrigation water and HYV seeds. It was observed that due to farness from city, the easy availability and distribution of agricultural inputs were relatively poor in distant UF. But despite farness from the city, farmers of distant UF were growing crops which were remunerative and commercial in nature. But these crops were not directly meant for city demands and being disposed off to either whole seller at village site itself or at agricultural mandis.

Table 4.8 Aligarh urban fringe: provisions of agricultural inputs

Zone		Provisions of agricultural inputs				
		Irrigation	HYV Seeds	Chemical fertilizers	Pesticides	Mechanical tools and equipments
Immediate urban fringe	Count	**181**	**165**	**157**	**41**	**116**
	% within zone	<u>100.00 %</u>	<u>91.20 %</u>	<u>86.70 %</u>	<u>22.70 %</u>	<u>64.10 %</u>
	% within fringe	41.80 %	41.50 %	41.50 %	51.30 %	36.40 %
Intermediate urban fringe	Count	**128**	**118**	**114**	**22**	**103**
	% within zone	<u>94.10 %</u>	<u>86.80 %</u>	<u>83.80 %</u>	<u>16.20 %</u>	<u>75.70 %</u>
	% within fringe	29.50 %	29.60 %	30.20 %	27.50 %	32.30 %
Distant urban fringe	Count	**124**	**115**	**107**	**17**	**100**
	% within zone	<u>88.57 %</u>	<u>82.10 %</u>	<u>76.40 %</u>	<u>12.10 %</u>	<u>71.40 %</u>
	% within fringe	28.60 %	28.90 %	28.30 %	21.50 %	31.30 %
Total	Count	**433**	**398**	**378**	**80**	**319**
	% within zone	<u>94.70 %</u>	<u>87.10 %</u>	<u>82.70 %</u>	<u>17.50 %</u>	<u>69.80 %</u>
	% within fringe	100.00 %	100.00 %	100.00 %	100.00 %	100.00 %

Notes
i. Figures in bold are absolute number of sampled respondents in the particular distance zones
ii. Figures in underlined are percentages to the total from that particular distance zone
iii. Figures below the underlined figures are percentages to the total of that particular variable
Source Based on field survey (2013)

4.9.4 Dynamics in Area Under Major Crops

To identify variation observed in the different zones in terms of land under different crops. In this case, one-way ANOVA with post hoc tests were applied. The test was done to examine difference in the mean of land under particular crop in different zones. Land under particular crop was taken as dependent variable while different urban fringe zone was assigned as independent variable.

One-way between-groups ANOVA test corroborates the survey results that there is a significant difference between three urban fringes in terms of land used for all the crops except oilseeds (Table 4.10). Significant value are 0.000, 0.000, 0.002, 0.006 for cereal, pulses, vegetables and other crops, respectively, which are significant at 1 % level. The statistical significance of the differences between each pair of groups is provided in the next Table 4.11, on Multiple Comparisons, which gives the result of the post hoc tests.

Table 4.9 Aligarh urban fringe: area under major crops in the farms of sampled households

Zone		Land under different crops				
		Cereals	Pulse	Oilseeds	Vegetables	Others
Immediate urban fringe	Count	**224**	**67**	**99**	**185**	**86**
	% within zone	33.9 %	10.1 %	14.9 %	27.9 %	13.0 %
	% within fringe	14.3 %	16.8 %	26.0 %	53.0 %	57.7 %
Intermediate urban fringe	Count	**463**	**124**	**128**	**139**	**36**
	% within zone	52.0 %	13.9 %	14.4 %	15.6 %	4.0 %
	% within fringe	29.6 %	31.2 %	33.6 %	39.8 %	24.2 %
Distant urban fringe	Count	**875**	**206**	**154**	**25**	**27**
	% within zone	67.9 %	16.0 %	11.9 %	1.9 %	2.1 %
	% within fringe	56.0 %	51.9 %	40.4 %	7.1 %	18.1 %
Total	Count	**1562**	**397**	**381**	**349**	**149**
	% within zone	55.0 %	13.9 %	13.4 %	12.3 %	5.2 %
	% within fringe	100.00 %	100.00 %	100.00 %	100.00 %	100.00 %

Notes
i. Figures in bold are agricultural land of sampled respondents in the particular distance zones
ii. Figures in underlined are percentages to the total from that particular distance zone
iii. Figures below the underlined figures are percentages to the total of that particular variable
Source Based on field survey (2013)

ANOVA test has indicated that there is a significant difference across the zones. The post hoc test result shows that immediate urban fringe is statistically different from the distant urban fringe in all crops except oilseeds, and intermediate urban fringe is statistically different from the distant urban fringe in cereal and other crop. Immediate urban fringe differs from distant urban fringe in all cases at the 1 % level of significance ($P < 0.01$) and intermediate urban fringe differs from distant urban fringe in both cases at 5 % level of significance ($p < 0.05$).

Table 4.10 Dynamics in area under major crops

Crops	Test of homogeneity of variances				ANOVA					
	Levene statistic	df1	df2	Sig.		Sum of squares	df	Mean square	F	Sig.
Cereals	34.085	2	454	0.090	Between groups	4845.38	2	2422.69	9.184	**0.000**
					Within groups	119764.62	454	263.79	–	–
					Total	124610.02	456	–	–	–
Pulse	41.901	2	454	0.120	Between groups	50.758	2	25.379	10.461	**0.000**
					Within groups	1101.45	454	2.426	–	–
					Total	1152.21	456	–	–	–
Oil seeds	1.556	2	454	0.212	Between groups	638.28	2	319.140	2.271	**0.104**
					Within groups	63799.15	454	140.527	–	–
					Total	64437.43	456	–	–	–
Vegetables	9.269	2	454	0.082	Between groups	649.871	2	324.94	6.486	**0.002**
					Within groups	22743.80	454	50.096	–	–
					Total	23393.672	456	–	–	–
Other crops	6.097	2	454	0.251	Between groups	42.529	2	21.264	5.199	**0.006**
					Within groups	1856.968	454	–	–	–
					Total	1899.497	456	–	–	–

*The mean difference is significant at the 0.01 level
N-457, Dependent variable-land under particular crop, Independent variables-zones (3 categories)

Table 4.11 Area under major crops: multiple comparisons

Crops	Area under different crops		Mean difference	Std. error	Sig.	95 % Confidence interval	
	Zone					Lower bound	Upper bound
Cereals	Immediate	Intermediate	−2.338	1.843	0.414	−6.672	1.995
		Distant	**−7.750***	1.828	**0.000**	−12.048	−3.451
	Intermediate	Immediate	2.338	1.843	0.414	−1.9955	6.672
		Distant	**−5.412***	1.955	**0.016**	−10.010	−0.813
	Distant	Immediate	**7.750***	1.828	**0.000**	3.451	12.048
		Intermediate	**5.412***	1.955	**0.016**	0.813	10.010
Pulses	Immediate	Intermediate	−0.408	0.176	0.056	−0.823	0.007
		Distant	**−0.799***	0.175	**0.000**	−1.211	−0.387
	Intermediate	Immediate	0.408	0.176	0.056	−0.007	0.823
		Distant	−0.391	0.187	0.094	−0.832	0.049
	Distant	Immediate	**0.799***	0.175	**0.000**	0.387	1.211
		Intermediate	0.391	0.187	0.094	−0.049	0.832
Oilseeds	Immediate	Intermediate	2.119	1.345	0.257	−1.043	5.283
		Distant	2.635	1.334	0.119	−0.501	5.772
	Intermediate	Immediate	−2.119	1.345	0.257	−5.283	1.043
		Distant	0.515	1.427	0.931	−2.840	3.871
	Distant	Immediate	−2.635	1.334	0.119	−5.772	0.501
		Intermediate	−0.515	1.427	0.931	−3.871	2.840
Vegetables	Immediate	Intermediate	1.736	0.803	0.079	−0.152	3.624
		Distant	**2.814***	0.796	**0.001**	0.9409	4.687
	Intermediate	Immediate	−1.736	0.803	0.079	−3.624	0.152
		Distant	1.077	0.852	0.416	−.925	3.081
	Distant	Immediate	**−2.814***	0.796	**0.001**	−4.687	−0.940
		Intermediate	−1.077	0.852	0.416	−3.081	0.925
Other crops	Immediate	Intermediate	−2.338	1.843	0.414	−6.672	1.995
		Distant	**−7.750***	1.828	**0.000**	−12.048	−3.451
	Intermediate	Immediate	2.338	1.843	0.414	−1.995	6.672
		Distant	**−5.411***	1.955	**0.016**	−10.010	−0.813
	Distant	Immediate	**7.750***	1.828	**0.000**	3.451	12.048
		Intermediate	**5.411***	1.955	**0.016**	0.813	10.010

*The mean difference is significant at the 0.05 level.
N-457, Dependent variable-land under particular crop, independent variables-zones (3 categories)

Wheat is the dominant crop, it's a cereal crop but because of high yield and assured market it is most favoured crop.

Mustard farming

Aligarh Urban Fringe: Vegetable farming has good market

Aligarh Urban Fringe: The urban demand for fruits

Papaya are locally produced

Oranges are quite popular in the region

Watermelon are locally produced

Watermelon are transported to other places

Fruits are marred by malpractices as chemicals are used for preserving and early ripening of fruits

References

Brook RM et al (eds) (2000) The peri-urban interface: a tale of two cities. University of Wales and Development Planning Unit, University College London, London

Chanda R (2002) Globalisation of services: india's opportunities and constraints. Oxford Press, Delhi

Coche AG (1982) Cage culture of tilapias. In: Pullin RSV, Lowe-McConnell RH (eds) Biology and culture of tilapias, ICLARM, Philippines, pp 205–246

Keivani R, Mattingly M (2007) The interface of globalization and peripheral land in the cities of the south: implications for urban governance and local economic development. Int J Urban Reg Res 31(2):462

Mandere NM et al (2010) Peri-urban development, livelihood change and household income: a case study of peri-urban Nyahururu, Kenya. J Agric Extension Rural Dev 2(5):78–79
Mehta R (2009) Rural livelihood diversification and its measurement issues: focus India. In: Wye city group on statistics on rural development and agriculture household income, Second meeting, FAO Head-Quarters, Italy, Rome, 11–12 June 2009. www.fao.org/.../rural/.../Paper_1_a1_Mehta-Wye-Grp-RurDiv-Final.doc
Prain G (2006) Participatory technology development for sustainable intensification of urban agriculture. In: Veenhuizen RV (ed) Cities farming for the future: urban agriculture for green and productive cities, RUAF Foundation, IDRC and IIRR Publishing, Philippines, pp 273–312

Chapter 5
Urban Fringe: Dairying Activity as Livelihood Option

The livestock production and agriculture are intrinsically linked to each other. India traditionally has agriculture-based rural economy which holds about 17 % share in nation's GDP. Dairying is a secondary occupation for about 69 % of India's farming community. Moreover, animal husbandry and dairy sector contribute over 31 % to the output of agriculture and contribute more than 5 % to nation's GDP. It has been predicted that this sector will play an increasingly important role in the national economy in the future. It forms an important livelihood activity for most of the farmers, supporting agriculture in the form of critical inputs, contributing to the health and nutrition of the household, supplementing incomes, offering employment opportunities and finally being a dependable "bank on hooves" in times of need (Government of India, Planning Commission 2007). The growth in agriculture sector over the years has been slow and fluctuating while livestock sector on the contrary has shown a steady growth and thus providing stability to the overall family income. In livestock sector, dairy is a high-growth sector and get reflected in the growing importance in contribution of these in the livestock economy. Dairying has become an important secondary source of income for millions of rural–urban families and has important role in providing employment as well as income generating opportunities. Livestock Sector not only provides essential proteins and nutritious human diet, but also plays an important role in utilization of non-edible agricultural by-products. According to estimates of the Central Statistical Organization (CSO), milk accounted for 68 % of the total value of output from livestock. India is one of the largest producers of milk in world, producing more than 130 million tonnes annually. In terms of output, milk is now the single largest agricultural commodity in India. The contribution of milk alone (Rs. 262,214.51 crore) was higher than paddy (Rs. 151,634 crore), wheat (Rs. 99,667 crore) and sugarcane (Rs. 58,470 crore) during 2010–2011.

N. Banu and S. Fazal, *Livelihood and Wellbeing in the Urban Fringe*,
The Urban Book Series, DOI 10.1007/978-3-319-39660-6_5

Continuous rise in per capita income and urbanization are fuelling rapid growth in demand for animal food products (Government of India, Planning Commission 2007). Recently the main focus of agricultural development initiatives has been on rural areas with the view that improved food production in rural areas can supply the expanding urban population. This is especially true for livestock production which has received little attention from research and development initiatives. Moreover peri-urban and urban livestock systems can play an important role to meet the growing needs for dairy commodities.

Urban fringe agriculture (UFA) is recognized not to be recent phenomena; however, its importance and potential for urban livelihoods has been neglected in the past. FAO in 1999 estimated that 800 million urban residents worldwide are engaged in UFA. The status of UFA has changed from illegal to tolerate and only in recent years UFA has begun to receive attention from donors, researchers and development organizations as a "new" development strategy. Most of the existing UFA studies have focused on crop production in urban and urban fringe areas, neglecting the role of livestock production.

Urban fringe are characterized by change, often rapid, resulting from urban development and expansion. This means appropriate strategies and efforts are required to utilize the opportunities for new livelihoods and also to thwart threats to existing livelihoods. Well planned and targeted interventions are required to enable people to make the best of new opportunities and overcome the threats. This requires understanding of the interactions between urban and rural areas characterizing the urban fringe. One emerging livelihood option, which needs thorough understanding, is the livestock farming in the urban fringe area. This activity appears to have great potential, but in India it is found to be poorly managed.

The previous chapter, focused on traditional methods of livelihood among sampled households in Aligarh's urban fringe. In this chapter the focus is on modifications and changes to traditional livelihood practices because of rapidly transforming urban fringe's economic environment characterized by monetized urban economy. The study assumes that the livestock-based livelihoods are greatly influenced by urban growth. The study makes an attempt to understand the status of the livestock farming system in the urban fringe driven by urban development by analyzing sampled households in Aligarh's urban fringe. This study seeks to present the findings gathered through household field survey as well as interviews with individuals like milk man, veterinary doctor, dairy owners, milk vendors who are the stake holders in dairy activity. Here descriptive statistics, non-parametrical test (Chi-squrae) and parametrical test (One-way ANOVA with Post Hoc test) are used to show how the city proximity affects dairy activity and how the households of different zone adapted it as their convenience.

Aligarh Urban Fringe: Dairy activity is a household affair

Livestock in urban fringe has great potential because of the demand for milk and dairy products which increases with growing urban population. Since most urban populations consist of 'non-productive consumers', the increase in demand has to be met by increased production from urban fringe. The bigger cities have large organized dairies procuring and supplying milk from urban fringe. But smaller cities have predominantly unorganized individual dairy producers in the urban fringe.

In India, urban populations, being diverse, also have diverse tastes as well as diverse demands for dairy products. One such demand is for fresh milk supplied directly by the producers and vendors at consumer's door steps. The other is the preference of Indian taste buds for buffalo milk. The urban milk consumers also have diverse preferences for the milk price that they would like to pay. On the basis of price paid, water is added to milk, which is an indigenous way of maintaining low or high fat content in milk as well as overall milk quality.

In and around the city there are many large and small dairies. About 30 commercial enterprises deal in milk trade and distribution. They are not the direct producers but involved in collecting, packaging and distributing of milk and its byproducts. While a number (more than 100) of smaller dairies, which are producers as well as at local level distributors of milk own up to 20 buffaloes and crossbreed cows. By far, the largest number of these smaller dairies belongs to traditional buffalo keepers, known as *gwalas*. Some of these households rely solely on the milk produced by buffaloes as their source of income, others rely more on urban-based work, but keep one or two buffaloes as a source of milk for their family and as an additional source of income. Keeping buffaloes is also part of their tradition. Animals may be impounded in a *baada or go-shaala or a gher*, which are cattle shelters.

5.1 Status of Livestock Farming

In urban fringe systems, households are found to engage in livestock production to supplement their incomes. Earlier very little attention was given to these systems, but they appear to be a growing phenomenon. At least in urban fringe, this is substantiated by several studies in developing countries which reported of emerging peri-urban livestock sector (Brookm 2005; Tacoli 1998; Fazal 2008). This livelihood option has inherent urban fringe characteristics as households adopt multiple options where the men can work in the farm as well as off-farm (in some urban activity or as middlemen), similarly, women can perform the men's as well as their own tasks of raising animals. So, dairying can be a full time job or can be performed as subsidiary activity.

5.2 Dairying as an Emerging Livelihood Option

In general, buffalo is the preferred milking animal in the urban fringe of Aligarh city, as consumers prefer its milk because of its high fat content. Between 1972 and 1982, the number of working cattle males declined sharply by 25 % (12 million in number) in India (Government of India, Planning Commission 2007). Out of the total buffalo population in the country, more than 50 % are breedable females, indicating preference of farmers for rearing buffaloes for milk production (ibid.). However, cows are also being domesticated by fringe households, though in lesser numbers. Cows are used basically to produce bullocks for agricultural sector and its milk is also produced for self-consumption as well as for specific demands of individuals (mainly for feeding infants because this milk have medicinal value and contains less fat). But with mechanization of agricultural sector and preference to buffalo for dairying activity in urban fringe, there is increase in the number of local buffaloes while a decrease in number of local cows over time (Table 5.1).

Table 5.1 Aligarh urban fringe: domesticated animals in sampled households

Zone		Cattle			Total
		Buffaloes	Cows	Goats	
Immediate urban fringe	Count	**538**	**113**	**52**	**703**
	% within zone	76.53	16.07	7.40	100.0
	% within fringe	47.28	45.93	68.42	48.15
Intermediate urban fringe	Count	**296**	**83**	**24**	**403**
	% within zone	73.45	20.60	5.96	100.0
	% within fringe	26.01	33.74	31.58	27.60
Distant urban fringe	Count	**304**	**50**	**0**	**354**
	% within zone	85.88	14.12	0.00	100.0
	% within fringe	26.71	20.33	0.00	24.25
Total	Count	**1138**	**246**	**76**	**1460**
	% within zone	77.95	16.85	5.21	100.0
	% within fringe	100.0	100.0	100.0	100.0

Notes
i. Figures in bold are absolute number of sampled respondents in the particular distance zones
ii. Figures in underlined are percentages to the total from that particular distance zone
iii. Figures below the underlined figures are percentages to the total of that particular variable
Source Based on field survey (2013)

5.3 Dairying Activity as Livelihood Option

The dairying activity has emerged as an important activity in Aligarh's urban fringe as it was discussed in Chap. 4 that out of the total sampled household 8.2 % have reported to have adopted it as the main livelihood option (Table 4.3). It is adopted as main livelihood by the fringe people because there is growing demands for milk as well as dairy products in the city and to meet this demand, the fringe people take it as livelihood option which fetches regular income. The survey shows that near about three-fourth households are practicing dairying activities for commercial purposes and rest one-fourth household produced milk for self-consumption (Table 5.2). The sample survey shows that dairy producers in immediate urban fringe are commercial producers of dairy products. Majority of immediate urban fringe (81 %) households are disposing off their dairy products to the market. But with distance from city, the share of self-consumption of dairy commodities increases and lesser amount of dairy production is marketed.

Out of 490 households associated with dairy activity, only 113 (23 %) pursued dairy activity for period of more than 10 years, while remaining 377 (77 %) have started to perform it in the past 10 years. This suggests that the growing demand for dairy products in urban areas and to meet this demand, more and more households are opting for dairying activity in urban fringe. The popularization of this activity is also because of cash remittance of the product. Significantly, 57 % households have

Table 5.2 Aligarh urban fringe: purpose of dairying activity of sampled households

Zone		Use of produced milk		Total
		Self purpose	Marketing purpose	
Immediate urban fringe	Count	**42**	**184**	**226**
	% within zone	18.6	81.4	100.0
	% within fringe	31.1	51.8	46.1
Intermediate urban fringe	Count	**56**	**86**	**142**
	% within zone	39.4	60.6	100.0
	% within fringe	41.5	24.2	29.0
Distant urban fringe	Count	**37**	**85**	**122**
	% within zone	30.3	69.7	100.0
	% within fringe	27.4	23.9	24.9
Total	Count	**135**	**355**	**490**
	% within zone	27.6	72.4	100.0
	% within fringe	100.0	100.0	100.0

Notes
i. Figures in bold are absolute number of sampled respondents in the particular distance zones
ii. Figures in underlined are percentages to the total from that particular distance zone
iii. Figures below the underlined figures are percentages to the total of that particular variable
Source Based on field survey (2013)

less than 2 cattles per households. This suggests that livestock rearing for producing dairy products in Aligarh's urban fringe was not the primary occupation for the rural population, but it only serves as a support enterprise to agriculture. Buffaloes are reared primarily for milk, butter, ghee and other milk products as well as for animal wastes (used as manure and fuel). Buffaloes are also preferred because consumers prefer its high fat content and as a result the milk fetches a better price.

5.4 Dairying Activity Is Thriving in Immediate Urban Fringe

Livestock rearing is gaining importance in urban fringe villages with increasing number of households involved over time. A total of 490 households were associated with dairy activity in Aligarh's urban fringe. Almost half of it (226 households), were found to be in *immediate* urban fringe, followed by *intermediate* urban fringe (142 households) and *distant* urban fringe (122 households) (Table 5.3). This clearly suggests that distance from urban area is critical for dairy activity. This is because milk being a perishable commodity and with no refrigeration or preservation techniques are used, this activity thrives closer to the city. Although more than half of the total sampled households in each urban fringe associated with dairy

Table 5.3 Aligarh urban fringe: dairying activity among sampled households

Zone		Households engaged in dairying activity		Total
		Yes	No	
Immediate urban fringe	Count	**226**	**167**	**393**
	% within zone	57.5	42.5	100.0
	% within fringe	46.1	62.5	51.9
Intermediate urban fringe	Count	**142**	**56**	**198**
	% within zone	71.7	28.3	100.0
	% within fringe	29.0	21.0	26.2
Distant urban fringe	Count	**122**	**44**	**166**
	% within zone	73.5	26.5	100.0
	% within fringe	24.9	16.5	21.9
Total	Count	**490**	**267**	**757**
	% within zone	64.7	35.3	100.0
	% within fringe	100.0	100.0	100.0

Notes
i. Figures in bold are absolute number of sampled respondents in the particular distance zones
ii. Figures in underlined are percentages to the total from that particular distance zone
iii. Figures below the underlined figures are percentages to the total of that particular variable
Source Based on field survey (2013)

activity, still there is significant difference between the different zones. A chi-square test for independence (Pearson Chi-square value) indicate significant association between distance from the city center and household's association with dairying activity, $\chi 2$ (1, n = 757) = 18.80, p = 0.00, Cramer's V = 0.16. Out of the total households in the urban fringe associated with dairying activity, almost half (46 %) belong to the immediate urban fringe and with increasing distance it decreases (Appendix E).

In immediate urban fringe, mean number of buffaloes per family was four times higher than the number in distant urban fringe. In the past five year only the number of buffaloes increased by 200 % in immediate urban fringe and only marginally in distant urban fringe. This is an indication of farmers investing in buffaloes as specialized dairy animals. This is particularly so in households near the urban centres where access to markets is better and a high demand for buffalo milk prevails. There are, however, examples of households reporting dairying activity from distant urban fringe also but these had better accessibility and well connected with Aligarh city by metalled roads, helping quick and easy transportation of dairy output.

Improved cow shed with proper water supply through bore wells

The dairy producers mostly sell their products in the city. But distance is a critical factor. Some producers sell from their own outlet.

5.5 Dairying Activity is Popular among Joint Family

Family size has great impact on dairy activity. The sample survey of Aligarh's urban fringe reveals that dairy activity was prominent among joint families or among households with larger family size. The livestock farming requires more labours and it becomes easy for large family members to look after the cattle. From the sample survey, it was found that out of the total households in each medium and large family type, 69 and 79 %, respectively, are engaged in livestock activity while out of the total small family households only 31 % households are associated with it (Table 5.4). Out of the total dairy households (490), 448 (91 %) households are either medium or large family type. The rest 42 households are small family type. This trend was also verified statistically. A chi-square test for independence (Pearson Chi-square value) indicated significant association between household's association with dairying and their family size, {$\chi 2$ (1, n = 757) = 82.60, p = 0.00, Cramer's V = 0.33}. The strong association between larger family size and dairying activity, this is mainly because operations in dairy activity requires labour for almost all round the day and family members are best suited for required operations (Appendix F).

Table 5.4 Aligarh urban fringe: family size and dairying activity among sampled households

Zone		Households engaged in dairying activity		Total
		Yes	No	
Small (less than 5 members)	Count	**42**	**91**	**133**
	% within zone	<u>31.6</u>	<u>68.4</u>	<u>100.0</u>
	% within fringe	8.6	34.1	17.6
Medium (5–8 members)	Count	**326**	**144**	**470**
	% within zone	<u>69.4</u>	<u>30.6</u>	<u>100.0</u>
	% within fringe	66.5	53.9	62.1
Large (more than 8 members)	Count	**122**	**32**	**154**
	% within zone	<u>79.2</u>	<u>20.8</u>	<u>100.0</u>
	% within fringe	24.9	12.0	20.3
Total	Count	**490**	**267**	**757**
	% within zone	<u>64.7</u>	<u>35.3</u>	<u>100.0</u>
	% within fringe	100.0	100.0	100.0

Notes
i. Figures in bold are absolute number of sampled respondents in the particular distance zones
ii. Figures in underlined are percentages to the total from that particular distance zone
iii. Figures below the underlined figures are percentages to the total of that particular variable
Source Based on field survey (2013)

5.6 Management in Dairying Activity

The study of management of the animals reveals a clear division of labour among the family members. The male members look after mainly the marketing affairs of the dairying activity, while female members take care of house affairs of dairying activity; this includes feeding, milking, and cleaning of animals. This division of labour is more prominent in distant urban fringe households, where animal husbandry is not the primary livelihood. As dairy has gained importance in the urban fringe economy over the past decade, the trend is changing with men being increasingly involved in animal management. This change is most evident in immediate urban fringe households where dairy has become an important livelihood activity. In the immediate urban fringe households, dairying has become a male dominated activity; women are confined with milking and cleaning work only. Except for the wealthier households, animal husbandry is a family activity where the family members do all management and very little hired labour is involved. The developments in this sector contributed to women empowerment in a big way. In the urban fringe of Aligarh city, women from each dairy household were found to be associated with dairying activity. The women play a significant role in animal husbandry and are directly involved in most of the operations relating to feeding, breeding, management and health-care of the livestock.

Aligarh Urban Fringe: Females also contribute actively in dairy farming

The male workers are engaged with other activities in agricultural activities or other non-farm activities, the female members of the households take care of the livestock within the house. Thus, the female members contribute substantially in dairy activity. Doing such a job in the household the female members also contribute for enhancing household's total income.

The survey shows that out of the total female workers in the urban fringe of Aligarh city 73.1 % engaged in primary economic activities with a high share of dairy farming with 45 % engaged in dairying activity (Fig. 5.1). The study also reveals the fact that there is difference in proportion of dairy workers, agricultural labourers and cultivators in the different zones of the urban fringe (Fig. 5.2). The

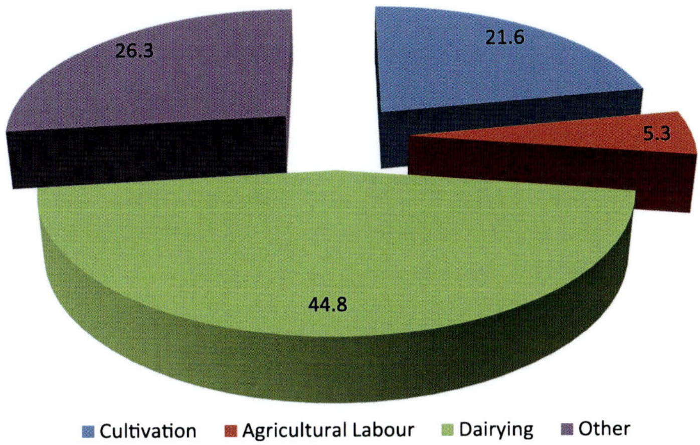

Fig. 5.1 Aligarh urban fringe: female's participation in dairying activity. *Source* Based on field survey (2013)

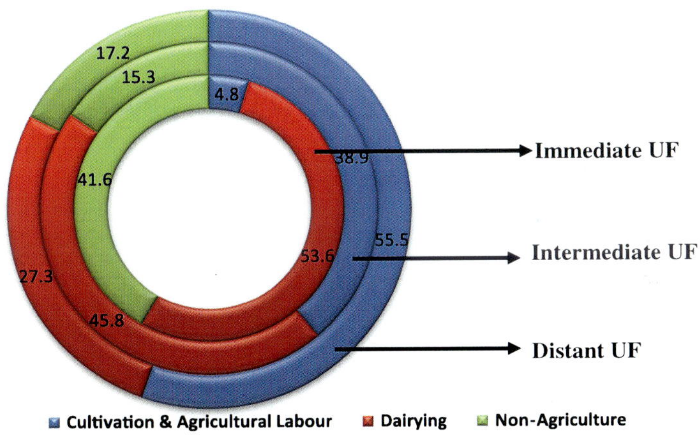

Fig. 5.2 Aligarh urban fringe: female's participation in different economic activities. *Source* Based on field survey (2013)

share of female workers associated with dairy activity in immediate urban fringe was 54 % which is followed by 46 and 27 % by intermediate urban fringe and distant urban fringe respectively. The trend was almost opposite for agricultural activities, i.e. cultivation and agricultural labourers, in which the distant urban fringe female workers (56 %) participation is higher than the immediate urban fringe (5 %) (Fig. 5.2).

Table 5.5 Aligarh urban fringe: cultivation and dairying activity among sampled households

Zone		Cultivation		Total
		Yes	No	
Immediate urban fringe	Count	**160**	**66**	**226**
	% within zone	70.8	29.2	100.0
	% within fringe	42.0	60.6	46.1
Intermediate urban fringe	Count	**107**	**35**	**142**
	% within zone	75.4	24.6	100.0
	% within fringe	28.1	32.1	29.0
Distant urban fringe	Count	**114**	**8**	**122**
	% within zone	93.4	6.6	100.0
	% within fringe	29.9	7.3	24.9
Total	Count	**381**	**109**	**490**
	% within zone	77.8	22.2	100.0
	% within fringe	100.0	100.0	100.0

Notes
i. Figures in bold are absolute number of sampled respondents in the particular distance zones
ii. Figures in underlined are percentages to the total from that particular distance zone
iii. Figures below the underlined figures are percentages to the total of that particular variable
Source Based on field survey (2013)

5.7 Dairying as a Supportive Livelihood Option

Even in the urban fringe of Aligarh city, households who engaged in dairy activity were also found to be associated with crop cultivation (Table 5.5). Livestock provides major additional contribution to agriculture through draft power, fuel, manure and as fertilizer. There is a strong relation between the dairy activity and crop cultivation. As discussed in Chap. 4 (Table 4.3) out of the 490 dairy producers, 381 are associated with crop cultivation also and another 62 households (13 % out of the total dairy producers) adopted dairying activity as main livelihood option because of its greater demand in the city and assured and regular income from it and rest 9 % households are engaged in non-farming activity along with dairying activity. Although increasing number of households are associating themselves with dairying activity but it is not the main livelihood option and for enhancing their income dairying has been taken as supplementary activity. Dairying is a secondary occupation for about 69 % of India's farming community (Government of India, Planning Commission 2007).

5.8 Household Consumption of Dairy Products

Traditionally dairying activity in subsistence society had majority production being consumed by the producers. But with evolvement of urban monetized economy, the share of household consumption is decreasing. Households in distant urban fringe

Table 5.6 Aligarh urban fringe: size of land holdings and household's consumption level

		Size of landholdings	Consumption of milk
Size of landholdings	Pearson correlation	1	**0.956**[a]
	Sig. (2-tailed)		0.000
	N	490	490
Consumption of milk	Pearson correlation	**0.956**[a]	1
	Sig. (2-tailed)	0.000	
	N	490	490

[a]Correlation is significant at the 0.01 level (2-tailed)
Source Based on field survey (2013)

households on an average consume more milk than those in immediate urban fringe households. This is presumably a function of urban demand and ease of access to urban market. The share of household consumption is very low in immediate fringe clearly pointing towards the urban demand. While its share increases with distance and distant fringe households' self consumption is higher. The consumption of milk has declined as the market for milk sale improved. When the milk consumption is averaged over all classes, it showed a decline during the past five years. The Pearson's correlation clearly shows that there is a strong positive correlation between the size of land and milk consumption at 1 % significant level (Table 5.6). Households with larger land holdings are able to consume more milk while marginal farmer's families or landless families have to sell as much as possible to maintain their livelihoods.

5.9 Milk Sales and Milk Marketing

Urban dwellers are the main consumers of milk and milk products produced in urban fringe households. Many of the producers (21 households) market milk door to door in the urban area. This is because of demand, higher rates, cash remittance, personal contact with customers (who sometimes help producers purchase animals by paying in advance) and the option to adjust quality of milk according to the price offered by the customer. Producers closer to the city can directly market their produce by-passing the middlemen, thus obtaining even higher revenues. Marketing of dairy products in Aligarh urban fringe have many channels (Fig. 5.3). The most common outlet is at their own houses as this is more prevalent in immediate urban fringe (Table 5.7). The consumers come down to households producing milk to get it as and when it is drawn. This way it is advantageous for both producers as well as consumers. Consumers get milk in pure form as when the product passes on through different hands the chances of its dilution by adding water increases. While producers through direct sale from their houses bypass the middleman or local traders and earn higher share of remuneration. However, the involvement of middlemen as local trades and city whole sellers is high in marketing of milk in Aligarh's urban fringe.

Table 5.7 Aligarh urban fringe: disposal of milk from the dairy households

Zone		Disposal of milk in the households					Total
		Not applicable	From own house	To local traders	Village dairy	Personally	
Immediate urban fringe	Count	**42**	**54**	**109**	**0**	**21**	**226**
	% within zone	18.6	23.9	48.2	0.0	9.3	100.0
	% within fringe	31.1	78.3	46.0	0.0	100.0	46.1
Intermediate urban fringe	Count	**56**	**13**	**47**	**26**	**0**	**142**
	% within zone	39.4	9.2	33.1	18.3	0.0	100.0
	% within fringe	41.5	18.8	19.8	92.9	0.0	29.0
Distant urban fringe	Count	**37**	**2**	**81**	**2**	**0**	**122**
	% within zone	30.3	1.6	66.4	1.6	0.0	100.0
	% within fringe	27.4	2.9	34.2	7.1	0.0	24.9
Total	Count	**135**	**69**	**237**	**28**	**21**	**490**
	% within zone	27.6	14.1	48.4	5.7	4.3	100.0
	% within fringe	100.0	100.0	100.0	100.0	100.0	100.0

Notes
i. Figures in bold are absolute number of sampled respondents in the particular distance zones
ii. Figures in underlined are percentages to the total from that particular distance zone
iii. Figures below the underlined figures are percentages to the total of that particular variable
Source Based on field survey (2013)

The sample survey shows that 48 % households dispose their milk to the local traders and it is more common in distant urban fringe where in 66 % households have to dispose it to the middlemen fringe (Table 5.7). Internal market is also found to be negligible in distant urban fringe and most of production comes to the city through traders. Chi-square test for independence (Pearson chi-square value) indicated significant association between distance from the city center and the type of disposal of produced milk in the fringe, $\chi 2$ (1, n = 490) = 140.99, p = 0.00, Cramer's V = 0.38. Out of the total households in the urban fringe associated with dairying activity, almost half (46 %) belong to the immediate urban fringe and with increasing distance it is decreasing. The immediate urban fringe dairy producers get more facility for disposal of milk because lesser distance from the city makes it easier for immediate urban fringe producers to visit the city personally to dispose their products. Along with the personal visit, disposal of milk from own households are more common in the closer zone than the distant urban fringe where the role of the middle men is prominent (Appendix G).

The average milk sales per household per year in immediate urban fringe are almost four times higher than that of household from distant urban fringe

Table 5.8 Aligarh urban fringe: disposal and consumption of milk among sampled dairy households

Zone	M (litre)	SC (litre)	TPM (litre)	No. dairy HH	PHS (litre)	PHC (litre)	PHP (litre)
Immediate urban fringe	3788	586	4374	226	16.76	2.59	19.35
Intermediate urban fringe	1397	662	2059	142	9.84	4.66	14.50
Distant urban fringe	591	747	1338	122	4.84	6.12	10.97
Total	5776	1995	7771	490	11.79	4.07	15.86

Source Based on field survey (2013)
M Marketed, *SC* Self consumption *TPM* Total produced milk
PHS Per household sale, *PHC* Per household consumption
PHP Per household productivity

(Table 5.8). The average income from these sales is even higher (close to six times). The survey reveals that the demand for dairy products especially milk is growing rapidly in Aligarh city. The supply of milk is not matching with the demand resulting in shortage of milk. This has resulted in rising prices and poor quality (addition of water to milk). Mean retail price of milk is higher for immediate urban fringe households (Rs. 30/l) compared to distant urban fringe households (Rs. 20/l). These mean values hide extremes. Cow's milk from distant urban fringe villages fetches a price as low as Rs 16/l while fresh buffalo milk sold in the city can sell for Rs. 40/l, a 3.3 fold difference.

To identify variations observed in the different zones in terms of average milk price, one-way ANOVA was applied. The test was done to examine the difference in the mean of milk price in different urban fringe zones. Milk price was taken as dependent variable while different urban fringe zone was assigned as independent variable. One-way between-groups ANOVA corroborate the survey results that there is a significant difference between three zones in terms of milk price (Table 5.9).

ANOVA test has indicated that there is a significant variations across the zones and the Post hoc test shows that the milk price in immediate urban fringe is statistically different from intermediate and as well as distant urban fringe at 5 % level of significance (Table 5.10). But between intermediate and distant urban fringe, the difference in average milk price was not significant. So it is clear that city proximity greatly influence the retail milk price which is to be more enjoyed by the immediate urban fringe than the other zones (Appendix H) (Table 5.11).

Table 5.9 ANOVA: mean retail price of produced milk

	Sum of squares	Df	Mean square	F	Sig.
Between groups	8173.961	2	4086.980	22.073	**0.000**
Within groups	90172.335	487	185.159		
Total	98346.296	489			

Source Based on field survey (2013)

Table 5.10 Tukey HSD test: mean retail price of produced milk

(I) Zone	(J) Zone	Mean difference (I − J)	Std. error	Sig.	95 % confidence interval	
					Lower bound	Upper bound
Immediate urban fringe	Intermediate urban fringe	**6.89524**[a]	1.45713	**0.000**	3.4697	10.3208
	Distant urban fringe	**9.29087**[a]	1.52872	**0.000**	5.6970	12.8847
Intermediate urban fringe	Immediate urban fringe	**−6.89524**[a]	1.45713	**0.000**	−10.3208	−3.4697
	Distant urban fringe	2.39564	1.67977	0.328	−1.5533	6.3446
Distant urban fringe	Immediate urban fringe	**−9.29087**[a]	1.52872	**0.000**	−12.8847	−5.6970
	Intermediate urban fringe	−2.39564	1.67977	0.328	−6.3446	1.5533

[a]The mean difference is significant at the 0.05 level
Source Based on field survey (2013)

Table 5.11 Aligarh urban fringe: prevailing retail price

Zone		Rate of milk per litre			Total
		Low rate (less than Rs. 25/-)	Medium rate (Rs. 25–35)	High rate (more than Rs. 40)	
Immediate urban fringe	Count	**123**	**36**	**67**	**226**
	% within zone	54.6	15.9	29.6	100.0
	% within rate of milk	42.4	27.1	100.0	46.1
Intermediate urban fringe	Count	**88**	**54**	**0**	**142**
	% within zone	61.9	38.0	0.0	100.0
	% within rate of milk	30.3	40.6	0.0	28.9
Distant urban fringe	Count	**79**	**43**	**0**	**122**
	% within zone	64.7	32.3	0.0	100.0
	% within rate of milk	27.2	6.8	0.0	24.9
Total	Count	**290**	**133**	**67**	**490**
	% within zone	59.1	27.1	13.6	100.0
	% within rate of milk	100.0	100.0	100.0	100.0

Notes
i. Figures in bold are absolute number of sampled respondents in the particular distance zones
ii. Figures in underlined are percentages to the total from that particular distance zone
iii. Figures below the underlined figures are percentages to the total of that particular variable
Source Based on field survey (2013)

Aligarh Urban Fringe: Most of the producers are in unorganized sector

Table 5.12 ANOVA: milk production

	Sum of squares	df	Mean square	F	Sig.
Between groups	940.879	2	470.440	7.857	**0.000**
Within groups	29,160.516	487	59.878		
Total	30,101.396	489			

To indentify the variations observed in milk productivity one-way between-groups ANOVA with post hoc test was applied. The test was done to examine the difference in the average milk productivity in different urban fringe zones. Milk production was taken as dependent variable while different urban fringe zone was assigned as independent variable. One-way between-groups ANOVA corroborates the survey results that there is a significant difference between three zones in terms of milk price (Table 5.12). ANOVA test has indicated that there is a significant variations across the zones and the Post hoc test shows that the milk production in immediate urban fringe is statistically different from intermediate and as well as distant urban fringe at 5 % level of significance (Table 5.13). But between intermediate and distant urban fringe, the difference in average milk productivity was not significant.

The sample survey also found that the households in distant urban fringe, who have lesser access to the urban market, prefer to convert the milk into milk products such as curd, *khoa* and *ghee* before selling them in the urban areas because the milk may get spoiled by the time they reach the city. Marketing of evening milk is also a problem from the distant urban fringe. Curd, *khoa* and *ghee* selling is a solution, but again these commodities are disposed off through middleman and thus the prices they get are lesser than the prevailing market price.

Table 5.13 Tukey HSD: milk production

Multiple comparisons						
(I) zone	(J) zone	Mean difference (I − J)	Std. error	Sig.	95 % confidence interval	
					Lower bound	Upper bound
(I) zone	(J) zone	**2.23345**[a]	0.82863	**0.020**	0.2854	4.1815
	Distant urban fringe	**3.21152**[a]	0.86934	**0.001**	1.1678	5.2552
Immediate urban fringe	Intermediate urban fringe	**−2.23345**[a]	0.82863	**0.020**	−4.1815	−0.2854
	Distant urban fringe	0.97807	0.95524	0.562	−1.2676	3.2237
Intermediate urban fringe	Immediate urban fringe	**−3.21152**[a]	0.86934	**0.001**	−5.2552	−1.1678
	Distant urban fringe	−0.97807	0.95524	0.562	−3.2237	1.2676

[a]The mean difference is significant at the 0.05 level

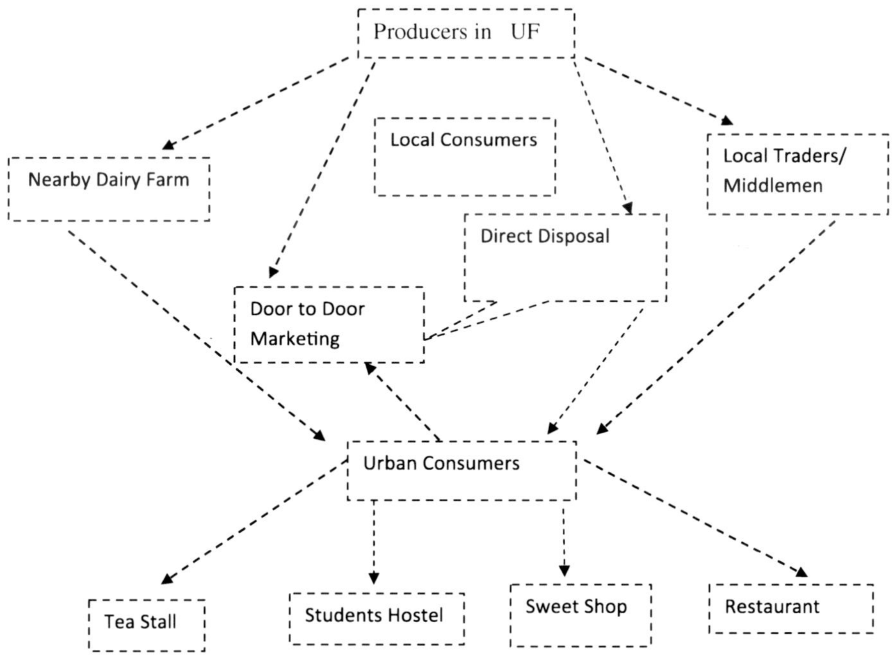

Fig. 5.3 Aligarh urban fringe: milk marketing channel

References

Brook R (2005) Enhancing livelihoods and natural resources management in Peri-urban villages. SAFS UW Bangor. (R8094 FTR)

Fazal S (2008) Farmers adaptation to urban forces: dynamics in the urban fringe of a growing urban centre in India. In: Bryant CR, Makhanya E, Herrmann TM (eds) The sustainability of rural systems in developing countries. Canada: Laboratoire de Développement durable et dynamique territoriale, Université de Montréal, pp 71–84

Government of India, Planning Commission (2007) Report of the working group on animal husbandry and dairying for the eleventh five year plan (2007–2012). New Delhi. file://\\cyber4\mtm\wrkg\WG-ON-ANIMAL-HUSBANDRY.htm

Tacoli C (1998) Rural-urban linkages and sustainable rural livelihoods. In: Carney D (ed) Sustainable rural livelihoods. DFID, London

Chapter 6
Urban Fringe: The Transformations

Transformation means change in appearance or nature in any phenomena. Here in this chapter the focus is on transformation in the character of urban fringe. Here we try to investigate and analyze the transformation of the rural settlements beyond the limits of a growing city (urban fringe) which underwent a process of change. This change is analyzed from two perspectives.

1. From the urban perspective, i.e. from the perspective of the people living in urban areas
2. From the rural perspective, i.e. from the perspective of the people living around the urban area—the urban fringe.

The economic development and the process of urbanization have resulted in urban fringe development. The particularity of its geographical location enables it to have two functions: urban landscape and rural landscape. The urbanization process is quite prominent among developing countries and with its development, in areas surrounding the cities (the urban fringe), the boundary between rural surrounding and city is becoming less and less distinct. Traditional isolated relationship between city and rural surrounding is now replaced by a new relationship that they rely on each other and penetrate into each other. This is the emergence of the urban fringe.

As in the previous chapters discussed, the formation and development in the urban fringe result from the interaction between rural surrounding and the city. The formation and development lead to the differentiation of dualistic structure of elements, functions and landscapes between rural and urban areas. Besides, the greater potential difference has brought about the interaction, communication and penetration between the two sides and enabled the evolvement of urban fringe. With the rural areas and urban areas penetrating into each other, a unique landscape pattern and a new way of utilizing land come into being, due to which a special environmental effect happens.

This chapter examines the transformations which have taken place in the urban fringe of Aligarh city through both primary and secondary sources of data. The secondary data was for village level amenities and facilities to assess changes at

N. Banu and S. Fazal, *Livelihood and Wellbeing in the Urban Fringe*,
The Urban Book Series, DOI 10.1007/978-3-319-39660-6_6

village level. These village level data were collected from District Statistical Bulletin (Uttar Pradesh) for 1995 and 2012. The transformations in urban fringe were further examined at household level in the sampled villages through field survey. The field survey was carried out to figure out the overall transformation in the urban fringes and to show the impact of distance from the urban centre on infrastructural facilities.

The functional force of city is the leadership mechanism of the formation of the urban fringe. The rapid development of city results also in flow of goods and capital, from urban areas to rural areas. This flow forms aerial expansion and immaterial element space spread. The spread of urban elements result from such factors as higher land price, higher tax, more serious pollution, heavier traffic and governments' policies about encouraging moving outside. The spread is influenced by the attenuation law of distance. The nearer an area is to the city, the more it will be influenced, the more urban elements it has. With the widened distance from the centre of the city, the urban elements in areas becomes less and less until they completely turn into rural elements. The space from urban elements to rural elements is the feature of urban fringe.

Similarly, the functional force of the rural area is the internal force of the formation of the urban fringe. The development of rural economy has influenced land utilization. Internal industry move outside to the boundary to make the formation of the urban fringe possible.

The development of the urban fringe is influenced not only by the radiation of the city centre and the inner strains of its urbanization, but also by the trend of the rural surrounding to aggregate to the city. The landscape elements of the urban fringe are mainly divided into two aspects.

The Rural Landscape Elements

Rural landscape elements consist of two parts: natural landscape elements, including farmlands, water bodies, orchards, etc., and human landscape elements, including villages, vernacular architecture, folk and traditional cultures, etc. Natural landscape elements are the backgrounds of the human landscape and the essential characteristics of rural settlements which are the carriers of the primary resource-based predominantly farming culture. This is characterized by traditional villages with green living places which are comfortable, peaceful, and simple and combines human, nature and society into one harmonious whole.

Rural landscape in the urban fringe is different from urban landscape and natural landscape. It is a complex constituted by settlement landscape, economic landscape, cultural landscape and environmental landscape. It differs from others, lies in its productive landscape and field utilization landscape as well as the unique rural culture and life.

The Urban Landscape Elements

Urban landscape typically takes human's dwell environment as its centre, including building, city industries, city amenities and facilities, streets, malls, multiplexes, parks and other recreational facilities, etc. All these factors have space-time nature and are complicated. Meanwhile, influenced by the self-protection consciousness of the communities, urban landscape elements are different in different regions. Functionally, urban landscape elements are presented in high-energy and high-capacity radiation spread as well as the cultural diversity. Urban landscape is different from the city centre to the boundary, with the transmutation of the intensity and patterns of communities, population density and functions.

The transformation has been witnessed largely in the following three broad areas, which were also interrelated with each other. They are:

(i) Transformations in economic environment of urban fringe
(ii) Transformations in social environment of urban fringe
(iii) Transformations in infrastructural facilities of urban fringe

6.1 The Transformation in the Economic Environment

The transformation in economic environment in the urban fringe is important simply because this zone is characterized by the coexistence of urban (mainly secondary and tertiary activities) and rural (mainly primary or natural resource-based) activities. Beyond towns, the benefits from economic growth have been distributed sparsely and radially dissipated (Fazal 2013). The urban fringes observe economic transformation in many ways. Some of them are discussed below.

6.1.1 Better Economic Opportunities

The households in Aligarh's urban fringe take advantage of the proximity to the city; the urban-based economic activities (consisting of secondary and rapidly expanding tertiary sector) draw large number of workers. The urban-based economic activities have employed about 29 % of the total sampled workers from the urban fringe. The proximity to the city was crucial as households from immediate urban fringe formed larger group of workers in these secondary and tertiary sectors. A major consequence of urbanization and industrialization is that a change is an increase in the spatial diversity of economic opportunity, the growth in tertiary employment and a reduction in the primary (Lewis and Maund 1976). Significantly, women from this zone also had some presence and they were also found to be taking advantage of available opportunities to work in city. Majority of these workers are lesser educated and do not possess skill or expertise. This is why these workers are engaged in lower level activities mostly in unorganized sector. But they have the advantage of drawing regular and assured monthly or daily salary as well as the nature of work also requires lesser labour. The survey shows that 37 %

workers from immediate urban fringe daily visit the city for earning their livelihood and with the increasing distance this trend decreases (Table 6.1). This suggests that the accessibility offers more economic opportunities for the closer zone as compared to the distant one.

6.1.2 Larger Market for the Fringe Producers

The urban fringe is also benefited from the larger urban market. The city with increasing population creates increased demand mainly for primary (nature-based) products which they produce in rural areas. The easy access (personal contact with consumers) and monetized urban economy is of great advantage to these urban fringe producers. The survey reveals that the urban fringe supplies a great variety of agricultural products (cereals, pulses, oilseeds, vegetables, fruits, flowers, etc.), animal products (meat, milk, ghee, curd, butter, etc.) and other natural products (earth, brick, wood, etc.).

The survey finds that an estimated 45,000 KL of milk, 425 quintals of vegetables, 200 quintals of fruits, 2000 quintals of chara and 700 quintals of cow dung are supplied to the city. As discussed in Chaps. 4 and 5, the survey reveals that both the cultivators as well as the dairy producers from the immediate urban fringe dispose their produced commodities in the city more than intermediate and distant urban fringes.

Table 6.1 Aligarh urban fringe: Place of work among sampled households

Zone		Place of work			Total
		Aligarh city	Fringe	Other	
Immediate urban fringe	Count	**303**	**445**	**82**	**830**
	% within zone	36.51 %	53.61 %	9.88 %	100.00 %
	% within fringe	61.59 %	43.16 %	45.56 %	48.74 %
Intermediate urban fringe	Count	**116**	**301**	**40**	**457**
	% within zone	25.38 %	65.86 %	8.75 %	100.00 %
	% within fringe	23.58 %	29.19 %	22.22 %	26.83 %
Distant urban fringe	Count	**73**	**285**	**58**	**416**
	% within zone	17.55 %	68.51 %	13.94 %	100.00 %
	% within fringe	14.84 %	27.64 %	32.22 %	24.43 %
Total	Count	**492**	**1031**	**180**	**1703**
	% within zone	28.89 %	60.54 %	10.57 %	100.00 %
	% within fringe	100.00 %	100.00 %	100.00 %	100.00 %

Notes
(i) Figures in bold are absolute number of sampled respondents in the particular distance zones
(ii) Figures underlined are percentages to the total from that particular distance zone
(iii) Figures below the underlined figures are percentages to the total of that particular variable
Source Based on field survey(2013)

This is because a relatively easy access to large urban market also helps the urban fringe producers to sell their products directly, manier times by passing the middlemen and draw higher revenues. While middlemen also play an active and dominant role (especially to household from distant urban fringe) in disposing commodities in urban market.

6.1.3 A Place Which Offers Monetized Economy

This feature of Aligarh urban fringe was quite evident in different distance zones. The respondents from immediate urban fringe have higher monthly income despite comparatively smaller family size while with increasing distance monthly income of households have declined (Table 6.2). The households have also showed association with diverse economic activity. This multi-activity household is survival response to urban monetized economy to generate higher incomes required for sustenance.

6.1.4 Diversification of Economic Activity

The urban fringe is influenced by urban monetized economy, larger market and greater opportunities. All this induce the urban fringe households to diversify their

Table 6.2 Aligarh urban fringe: Per capita income pattern among sampled households

Zone		Per capita monthly income			Total
		Low (less than Rs. 1300)	Medium (Rs. 1301 to Rs. 2600)	High (above Rs. 2600)	
Immediate urban fringe	Count	**103**	**115**	**175**	**393**
	% within zone	26.2 %	29.3 %	44.5 %	100.0 %
	% within fringe	40.4 %	50.4 %	63.9 %	51.9 %
Intermediate urban fringe	Count	**59**	**81**	**58**	**198**
	% within zone	29.8 %	40.9 %	29.3 %	100.0 %
	% within fringe	23.1 %	35.5 %	21.2 %	26.2 %
Distant urban fringe	Count	**93**	**32**	**41**	**166**
	% within zone	56.0 %	19.3 %	24.7 %	100.0 %
	% within fringe	36.5 %	14.0 %	15.0 %	21.9 %
Total	Count	**255**	**228**	**274**	**757**
	% within zone	33.7 %	30.1 %	36.2 %	100.0 %
	% within fringe	100.0 %	100.0 %	100.0 %	100.0 %

Notes
(i) Figures in bold are absolute number of sampled respondents in the particular distance zones
(ii) Figures underlined are percentages to the total from that particular distance zone
(iii) Figures below the underlined figures are percentages to the total of that particular variable
Source Based on field survey(2013)

livelihood for better and improved living. The study finds that increasing number of members in households diversify their sources of income and draw on a portfolio of activities, such as formal waged employment, informal trading and low-level services, etc. Large number of households reported that different members of the household are engaged in different types of activities and sectors of the economy.

The sample study finds that in the Aligarh's urban fringe, income diversification is quite common where increasing number of households are combining non-farming activity with their traditional farming activity. As discussed in Chap. 3, the survey shows that 68 % of the total sampled household's income is coming from more than one economic sector and this is more common in immediate urban fringe and decreases with decreasing accessibility to the city (Table 3.4). This diversification is necessary for enhancement in earnings required for survival in urban fringe. This argument is also strengthened by the fact that this trend is found to be more dominant in villages closer to the city. The survey shows that 45 % households in the immediate urban fringe are supplementing their income with adding dairying activity to their traditional farming activity. Combining farming and non-farming activities within one household plays a significant role in enhancing income as well as providing livelihood option which can be managed along with the farming activity.

The sample survey also finds that urban fringe household's women were also actively participating and contributing to household's income. Traditionally, these women contribute in farming and dairying activity but sample survey finds that women from immediate urban fringe are also engaged in urban-based non-farm activity. The survey also points towards the declining economic significance of farming activity as it is low income generating activity; farmers are diversifying to other non-farm activities for additional income. A good number of migrants to the urban fringe, especially those who do not have land sell their labour to non-farm sector while their other members work in farming sector in their native rural areas. This pattern of diversification observed in the urban fringe is essential because it reduces dependence on one source of livelihood as well as increases the income of household which reduces vulnerability to stresses and shock of livelihood in urban economy.

6.1.5 Higher and Regular Wages

Urban fringe not only offers greater economic opportunities but it also provides higher and assured wages. Traditionally, the wages in the rural agricultural sector are in kind rather than cash and that too is poor, irregular and incidence of exploitation and intimidation are common. The urban fringe with urban-based activities offers these workers an opportunity to sell their labour for remittance in cash. Urban fringe offers opportunities to all kinds of workers, it includes skilled and unskilled workers in secondary sector and trained and educated employed in tertiary sector. The overall wages for all kinds of urban-based activities are relatively higher and regular than in rural areas. This feature of urban-based activities attracts workers from the rural areas instigating migration.

6.1.6 Occupational Transformation

Another striking feature observed in Aligarh's urban fringe is declining preference to agriculture as livelihood option. This leads to large-scale occupational transformation in the urban fringe. This occupational transformation is influenced by popularization of education and this phenomenon is more prominent among younger generation. Moreover, this change has been clear urban fringe character becoming particularly strong with proximity to the city where greater choices and opportunities exist (Fig. 6.1). Another reason for this transformation is decreasing land for farming activity, which means small farmers do not have choice but to opt for urban-based livelihood, shedding their traditional farming activity (Table 6.3).

6.1.7 Increasing Land Value

High population densities on a limited area increase the pressure on land for food, space and leisure, and should in principle lead to higher food prices and higher incomes for farmers. Current demographic trends pose a dual land problem: scarcity and competition for land in the urban and urban fringe areas, and abandonment and lack of maintenance of property in remote rural areas with low carrying capacity and limited earnings from land use. Competition for available land in an open market system leads to an increase in land prices (FAO 2003). The urban fringe is

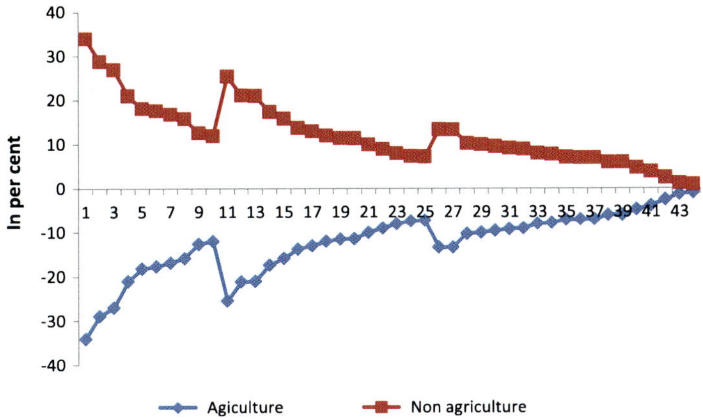

Fig. 6.1 Annual change in economic activities in urban fringe of Aligarh City from 1981–2001. *Source* Based on population census of India

Table 6.3 Aligarh urban fringe: Change in economic activity among sampled households

Zone		Change in economic activity			Total
		Primary	Secondary	Tertiary	
Immediate urban fringe	Count	**53**	**7**	**23**	**83**
	% within zone	63.85 %	8.43 %	27.71 %	(21.11 %)[a]
	% within fringe	55.78 %	87.50 %	76.66 %	62.40 %
Intermediate urban fringe	Count	**27**	**0**	**4**	**31**
	% within zone	87.09 %	0.00 %	12.90 %	(15.65 %)[a]
	% within fringe	28.42 %	0.00 %	13.33 %	23.31
Distant urban fringe	Count	**15**	**1**	**3**	**19**
	% within zone	78.94 %	5.26 %	15.78 %	(11.44 %)[a]
	% within fringe	15.78 %	12.50 %	10.00 %	14.29
Total	Count	**95**	**8**	**30**	**133**
	% within zone	71.42 %	6.01 %	22.55 %	
	% within fringe	100.00 %	100.00 %	100.00 %	

Notes
(i) Figures in bold are absolute number of sampled respondents in the particular distance zones
(ii) Figures underlined are percentages to the total from that particular distance zone
(iii) Figures below the underlined figures are percentages to the total of that particular variable
(iv) [a]Indicates the share to from that particular distance zone
Source Based on field survey(2013)

also characterized by increasing value of land. This trend has made land as an attractive investment object in urban fringe. This trend has affected the urban fringe economy in many ways. The outcome is felt in the form of large-scale as well as frequent land transformations. The agricultural land is the main sufferer which is shrinking and vacant land (the speculative land) is rapidly increasing. The use of land and its transformation in urban fringe is increasingly influenced by real estate developments which are instrumental in manipulating the prospects of future developments of land in urban fringe. The increase in land value is driven by increase in demand (the population pressure) but this increase in demand is also because of increase in land value. Land value and demand for land have bearing on each other. The increase in land value and increased demand for land in urban fringe has affected the agricultural system. The benefits of increases in land value have benefited more the large land owners while small land owners have been further marginalized. Village residents who live closer to towns can more easily benefit from these linear radial patterns of economic development, their lands grow in value and their labour is in greater demand. Those who live further out have comparatively little to offer that is of equal value. Their lands are not attractive to builders and industrialists. According to District Magistrate Office of Aligarh District, the land value in 2007 is 4.7 million per hectare, 2.7 million per ha and 1.3 million per ha in immediate, intermediate and distant urban fringes, respectively. This reveals the facts that closer zone are more in demand for the development purpose because of its easy accessibility.

6.2 The Transformation in Social Environment

The transformation in social environment in the urban fringe is important simply because this zone is characterized by the intermixing of urban way of life and rural way of living. This intermixing is in some way is constrained by cultural differences. Urbanization is one of the most pervasive processes of change shaping societies around the world (Bryant 1995). Increasing incomes particularly from the growing tertiary sector, and the possibility of upward social mobility have popularized education in the fringe. All the changes have transformed the social set up in the fringe. The urban fringe observes social transformation in many ways. Some of them are discussed below:

6.2.1 *Changing Demographic Features*

The urban fringe is witnessing rapid growth in its population as well as marked with intensification of concentration of population (Table 6.4). Gradient analysis lent considerable support to the identification of a peripheral rural–urban fringe. Basically, these models argue that population density declines with distance from the city centre and that a peripheral zone of high rural density can be envisaged (Lewis and Maund 1976). In Aligarh's urban fringe the population density has increased from 1765 per km^2 in 1995 to 2585 per km^2 in 2012. There is a continuous and accelerated inflow of population in immediate urban fringe, both from urban core and rural hinterland. The city population is moving to urban fringe because this zone provides less congested accommodation that too at lesser price whereas rural people migrate to urban fringe to take advantage of greater economic opportunities. Along with these facilities, relatively higher and regular wages act as pull factor for the migrants from far areas to the urban fringe especially in immediate urban fringe. The survey data clearly indicate that the concentration of migrants either from the city or nearby rural hinterland is only accumulated in the closest zone of the city, immediate urban fringe (Tables 6.5, 6.6 and 6.7).

6.2.2 *Conflicts Among Old Residents and Migrants*

The growth of new residential colonies within the urban fringe village dramatically alters sociological profile of the village (Ramachandran 1989). The old village settlement continues to exist almost intact with its original residents, while the new residential colonies house the people from the city. The social relationship between the old residents and migrants coming from the city is at best tenuous and

Table 6.4 Aligarh urban fringe: Population growth in the sampled villages

Zone		Population			Change	Change
		1991	2001	2011	1991–2001	2001–2011
Immediate urban fringe	Count	**14,766**	**26,329**	**48,979**	**78.31 %**	**86.03 %**
	% within fringe	31.39 %	41.12 %	51.51 %		
Intermediate urban fringe	Count	**16,898**	**20,527**	**25,086**	**21.48 %**	**22.21 %**
	% within fringe	35.92 %	32.06 %	26.10 %		
Distant urban fringe	Count	**15,381**	**17,168**	**21,020**	**11.62 %**	**22.43 %**
	% within fringe	32.69 %	26.81 %	22.10 %		
Total	Count	**47,045**	**64,024**	**95,085**	**36.09 %**	**48.51 %**
	% within fringe	100 %	1000 %	100 %		

Notes
(i) Figures in bold are absolute number of sampled respondents in the particular distance zones
(ii) Figures underlined are percentages to the total from that particular distance zone
Source Based on Population Census of India (1991, 2001, 2011)

superficial. The two communities have different perceptions of the city. However, there are some contacts between the two communities which are largely for their mutual benefits. The migrants remit their products for money which they could not get among their own social group. The supply of milk, vegetables and women are going to newly built colonies for domestic help.

Table 6.5 Aligarh urban fringe: Relocation of settlement among sampled households

Zone		Household Status		Total
		Native	Migrated	
Immediate urban fringe	Count	**268**	**125**	**393**
	% within zone	68.2 %	31.8 %	100.0 %
	% within fringe	42.4 %	100.0 %	51.9 %
Intermediate urban fringe	Count	**198**	**0**	**198**
	% within zone	100.0 %	0.0 %	100.0 %
	% within fringe	31.3 %	0.0 %	26.2 %
Distant urban fringe	Count	**166**	**0**	**166**
	% within zone	100.0 %	0.0 %	100.0 %
	% within fringe	26.3 %	0.0 %	21.9 %
Total	Count	**632**	**125**	**757**
	% within zone	83.5 %	16.5 %	100.0 %
	% within fringe	100.0 %	100.0 %	100.0 %

Notes
(i) Figures in bold are absolute number of sampled respondents in the particular distance zones
(ii) Figures underlined are percentages to the total from that particular distance zone
(iii) Figures below the underlined figures are percentages to the total of that particular variable
Source Based on field survey(2013)

Table 6.6 Aligarh urban fringe: Relocation period of sampled households

Zone		Relocation period			Total
		Less than 5 year	5–10 year	More than 10 year	
Immediate urban fringe	Count	**50**	**42**	**33**	**125**
	% within zone	40.0 %	33.6 %	26.4 %	100.0 %
	% within fringe	100.0 %	100.0 %	100.0 %	100.0 %
Total	Count	**50**	**42**	**33**	**125**
	% within zone	40.0 %	33.6 %	26.4 %	100.0 %
	% within fringe	100.0 %	100.0 %	100.0 %	100.0 %

Notes
(i) Figures in bold are absolute number of sampled respondents in the particular distance zones
(ii) Figures underlined are percentages to the total from that particular distance zone
(iii) Figures below the underlined figures are percentages to the total of that particular variable
Source Based on field survey(2013)

Table 6.7 Aligarh urban fringe: Place of origin among sampled households

Zone		Place of origin				Total
		Aligarh city	Nearby rural hinterland	Other district	Other state	
Immediate urban fringe	Count	**14**	**80**	**17**	**14**	**125**
	% within zone	11.2 %	64.0 %	13.6 %	11.2 %	100.0 %
	% within fringe	100.0 %	100.0 %	100.0 %	100.0 %	100.0 %
Total	Count	**14**	**80**	**17**	**14**	**125**
	% within zone	11.2 %	64.0 %	13.6 %	11.2 %	100.0 %
	% within fringe	100.0 %	100.0 %	100.0 %	100.0 %	100.0 %

Notes
(i) Figures in bold are absolute number of sampled respondents in the particular distance zones
(ii) Figures underlined are percentages to the total from that particular distance zone
(iii) Figures below the underlined figures are percentages to the total of that particular variable
Source Based on field survey (2013)

6.2.3 *Emergence of New Family System*

Urban fringe has also witnessed transformation in family structure. Household structure is an important factor. It is sometimes claim that urbanization leads to a preference for nuclear households instead of the joint households that are prevalent in agricultural households in rural India. The families are increasingly becoming nuclear, which includes married couple and their unmarried children. Traditionally, the households followed joint family norms where all the resources were pooled and elder member of the family distributes them among all the family members.

This system was suited to rural-based farming activity. But with changes in occupational structure, this family norm has also been changed. Ramachandran (1989) also claims that joint households come under pressure in semiurban society. Because of urban to rural migration, farmers increasingly have non-farming neighbours who unaccustomed to the sights, sounds and smells of modern agriculture (Diakosavvas 2008).

The sample survey finds that increase in land values have led to higher stakes as well as disputes among family members. Thus to avoid disputes, there are increasing number of families which now divide their lands leading to fragmentation of land. This division of land also results in division of economic dependence leading to nuclear family norm. The immediate urban fringe has more nuclear families because of more urban-based occupations and shortage of space where as distant urban fringe is still dominant of agricultural activity so still joint families exists(Table 6.8).

Table 6.8 Aligarh urban fringe: Type of family size of the sampled households

Zone		Family size			Total
		Small (less than 5 members)	Medium (5–8 members)	Large (more than 8 members)	
Immediate urban fringe	Count	**70**	**261**	**62**	**393**
	% within zone	17.8 %	66.4 %	15.8 %	100.0 %
	% within fringe	52.6 %	55.5 %	40.3 %	51.9 %
Intermediate urban fringe	Count	**42**	**116**	**40**	**198**
	% within zone	21.2 %	58.6 %	20.2 %	100.0 %
	% within fringe	31.6 %	24.7 %	26.0 %	26.2 %
Distant urban fringe	Count	**21**	**93**	**52**	**166**
	% within zone	12.7 %	56.0 %	31.3 %	100.0 %
	% within fringe	15.8 %	19.8 %	33.8 %	21.9 %
Total	Count	**133**	**470**	**154**	**757**
	% within zone	19.6 %	62.1 %	20.3 %	100.0 %
	% within fringe	100.0 %	100.0 %	100.0 %	100.0 %

Notes
(i) Figures in bold are absolute number of sampled respondents in the particular distance zones
(ii) Figures underlined are percentages to the total from that particular distance zone
(iii) Figures below the underlined figures are percentages to the total of that particular variable
Source Based on field survey(2013)

6.2.4 *Materialistic Attitude*

Urban fringe presents a meeting point of people from different walks of life having no affinity with one another. Everyone of urban fringe is rational regarding economic point of view because trend of materialism is continuously growing here. In materialistic society money is considered as the key to happiness. The sense of social responsibility among the people has been reduced. Competition for space and services in monetized economy has brought change in attitude. Owning agricultural land was once considered to enjoy higher social status but now jobs in tertiary sector are preferred. Interestingly, some respondents have blamed the popularization of satellite television among the rural masses. Increasing number of rural youth and females were found to be devoting their leisure time on watching daily soaps, serials and movies. This has made effects on their social life and impacts on family relations. In social terms an expanding and diversifying tertiary sector offers an increase in employment opportunities and creates new aspirations for more people (Saville 1957).

6.2.5 *Emergence of New Social Class*

The sampled households in Aligarh urban fringe represent a new social class which has increasing elements of urban characteristics and decreasing rural elements. The urban fringe elements households showed better education levels, especially younger generation had better educational attainments. They have smaller family size but marked gender biasness. Although they showed gender biasness among kids, status of women has improved especially with educational attainments among women and higher work participation. The households, where women working in tertiary activities attained higher status in the family as well as in the society. It clearly shows that economic independence is directly associated with women empowerment.

6.2.6 *Improving Women Status*

Another important social feature which was observed in the urban fringe was improving status of women. The urban fringe respondents have reported of better status of women than both rural interiors and urban core. The women were found to be actively participating in economic activities, managing the household as well as taking care of household resources too. The education, political empowerment and economically activeness have made them secure. There was male-specific migration in the immediate urban fringe resulting in somewhat skewed sex ratio, but still the overall status was satisfactory.

One important change which the urban fringe has witnessed in the urban fringe is awareness and popularization of education among the masses. Education has started to transform the urban fringe society. Even the poor and uneducated households have also started to push their children for pursuing education. Although, the government is injecting money and incentives but it has not made much impact. However, the social awareness has contributed significantly in improvement in education level. The respondents were found to prefer private schools only to ensure better and disciplined education. The government school education is greatly subsidized and there are provisions of other incentives like mid-day meal, school uniform, books and stationary, etc. But despite all, this poor management of schools and irregular teaching has distracted the parents.

6.2.7 Changing Social Perception

The sample survey finds that the sample households of Aligarh city have certainly benefited from urban proximity. They take advantage from the city's economic developments. The educational and health facilities of the city have greatly helped the urban fringe residents in the attainment of their educational and health requirements.

Overall also the city's infrastructure and facilities have helped in the development of urban fringe villages. However, the distance from the city was very crucial and obviously the urban fringe villages closer to the city drew more benefits than the distant villages in urban fringe. Interestingly when the respondents were enquired about the benefits from the city, the respondents from immediate and intermediate urban fringes were critical of the urban forces.

These respondents were concerned as their economic, social and cultural activities have been greatly transformed due to urban proximity. Which in their view have changed their livelihood and life style? But still these respondents do not want to shift their base, meaning which on one hand they want to take all the benefits of urban proximity while on other hand they want to remain stuck with their rural lifestyle and do not wish any erosion in social and cultural values. However, this dichotomy was not evident among younger generation. The distant urban fringe respondents on the other hand complain of difficulties in accessing urban benefits, they want rapid urban development and expansions so that they come closer to the city and get benefits of urban amenities and facilities.

6.2.8 *Improving Living Standards*

To assess the household quality of life only for the sampled villages in the urban fringe of Aligarh city, a Household Quality of Life Index (HQLI) has been made on the basis of 13 variables. HQLI is made up of two distinctive indices such as housing condition index (HCI) and asset possession index (API). Selected 13 variables are under these two indices. The variables details are given below.

Variables used for housing condition index	Variables used for asset possession index	Household quality of life index
• Bricked House • Drinking water within the premises • Electricity • Sanitation • Ventilation • LPG as fuel	• Television • Bicycle • Scooter/Motorcycle/Moped • Car/Jeep/Van • Refrigerator • Washing machine • Cooler	**Housing condition index + Asset possession index**

The HQLI is made with the help of range equalization (normalization) technique. Here, all unidirectional variables are normalized by dividing their range (maximum–minimum) to the difference from observed to minimum value (observed- minimum). Scale-free scores of observations of a variable vary from 'zero' to 'one'. After adding up all the scores of different variables of a particular observation create the index. Higher the score indicates the better living condition of the households.

The sample survey finds that there was a marked influence of city life on the life style and quality of life of urban fringe households. There were significant improvements in the urban fringe households which have become more urbanized and the quality of life has also improved considerably. Most of this improvement in urban fringe is because of inflow of money through sale of land or income from secondary and tertiary sectors. This improvement is visible from their houses (which are mostly pucca, constructed using bricks and cement). Distance from market, both physical and cognitive, significantly influence individuals, economic prospects. By alternatively facilitating or limiting access to market-based opportunities, distance can inverse critically between effort and reward.

Here have been improvements also in sanitation, ventilation, electricity connections, use of fuel, availability of modern gadgets, etc. However, all these improvements decrease with distance, as immediate urban fringe has better housing condition where as distant urban fringe has relatively poor condition (Table 6.9).

Table 6.9 Aligarh urban fringe: General status of living conditions among sampled households

Housing condition	Immediate urban fringe	Intermediate urban fringe	Distant urban fringe
Sanitation	**252**	**110**	**51**
	64.1 %	55.6 %	30.7 %
Ventilation	**284**	**106**	**50**
	72.3 %	53.5 %	30.1 %
Electricity	**364**	**143**	**95**
	92.6 %	72.2 %	57.2 %
Television	**342**	**144**	**83**
	87.0 %	72.7 %	50.0 %
Refrigerator	**212**	**59**	**45**
	53.9 %	29.8 %	27.1 %
Cooler	**266**	**77**	**57**
	67.7 %	38.9 %	34.3 %
Cycle	**204**	**79**	**13**
	51.9 %	39.9 %	7.8 %
Motor cycle	**256**	**89**	**61**
	65.1 %	44.9 %	36.7 %
Car	**47**	**12**	**7**
	12.0 %	6.1 %	4.2 %
Washing machine	**106**	**12**	**4**
	27.0 %	6.1 %	2.4 %

Notes
(i) Figures in bold are absolute number of sampled respondents in the particular distance zones
(ii) Figures underlined are percentages to the total from that particular distance zone
Source Based on field survey(2013)

The effects of the differences in earning potentials become visible when we look at asset ownership. The scores in API show the variations in each type of assets examined here which is owned by a greater proportion of households in the immediate urban fringe compared to the distanced zones reflecting the effects of distances in earning opportunities. Substantial differences in asset holdings are also visible within the urban fringe areas (Table 6.9). Every asset type examined here is possessed by larger share of households in immediate urban fringe as compared to distant urban fringe villages.

The effects of the differences in economic improvement become visible when we look at the HCI, API (Table 6.13) and in HQLI (Fig. 6.2). There have been improvements also in housing condition as well as in modern asset possession. However, these improvements increase with city accessibility, as immediate urban fringe has better housing condition and higher asset possession where as distant urban fringe has relatively poor housing condition and lower asset possession (Tables 6.9, 6.10, 6.11 and 6.12). In HQLI, 7 villages (64 %) out of 11 villages in the immediate urban fringe reported of having high quality of life and rest 4 villages (36 %) having medium quality of life. But as we move out to distant urban fringe 15 villages (83 %) out of 18 villages fall under low quality of life and rest 3 villages (17 %) having medium quality of life but not a single one reported in the category of high quality of life. This points towards impact of city proximity in the improvement on quality of life.

Aligarh Urban Fringe: Improvements in living conditions

Aligarh Urban Fringe: Houses have modern electrical and electronic gadgets of comfort

6.3 The Transformation of Infrastructural Facilities

The process of urbanization has its impact on its surrounding (the urban fringe) in many ways and at different scales. Urban impact on urban fringe is both in terms of 'site' and 'situation'. Here, 'site' is referred to the physical and cultural characteristics and attributes of the city. While, 'situation' specifies the external relations of the locale. In the context of an urban fringe, a change in situation usually

Table 6.10 Aligarh urban fringe: Type of house construction among sampled households

Zone		Type of house construction			Total
		Mud—Brick Thatched	Brick–Cement Non-RCC Columns–Roof	Brick–Cement with RCC Columns–Roof	
Immediate urban fringe	Count	**10**	**46**	**337**	**393**
	% within zone	2.5 %	11.7 %	85.8 %	100.0 %
	% within fringe	20.0 %	24.0 %	65.4 %	51.9 %
Intermediate urban fringe	Count	**25**	**72**	**101**	**198**
	% within zone	12.6 %	36.4 %	51.0 %	100.0 %
	% within fringe	50.0 %	37.5 %	19.6 %	26.2 %
Distant urban fringe	Count	**15**	**74**	**77**	**166**
	% within zone	9.0 %	44.6 %	46.4 %	100.0 %
	% within fringe	30.0 %	38.5 %	15.0 %	21.9 %
Total	Count	**50**	**192**	**515**	**757**
	% within zone	6.6 %	25.4 %	68.0 %	100.0 %
	% within fringe	100.0 %	100.0 %	100.0 %	100.0 %

Notes
(i) Figures in bold are absolute number of sampled respondents in the particular distance zones
(ii) Figures underlined are percentages to the total from that particular distance zone
(iii) Figures below the underlined figures are percentages to the total of that particular variable
Source Based on field survey (2013)

Table 6.11 Aligarh urban fringe: Use of fuel type among sampled households

Zone		What type of fuel is used			Total
		Wood/Cow dung	LPG	Both	
Immediate urban fringe	Count	**140**	**165**	**88**	**393**
	% within zone	35.6 %	42.0 %	22.4 %	100.0 %
	% within fringe	38.8 %	83.3 %	44.4 %	51.9 %
Intermediate urban fringe	Count	**101**	**27**	**70**	**198**
	% within zone	51.0 %	13.6 %	35.4 %	100.0 %
	% within fringe	28.0 %	13.6 %	35.4 %	26.2 %
Distant urban fringe	Count	**120**	**6**	**40**	**166**
	% within zone	72.3 %	3.6 %	24.1 %	100.0 %
	% within fringe	33.2 %	3.0 %	20.2 %	21.9 %
Total	Count	**361**	**198**	**198**	**757**
	% within zone	47.7 %	26.2 %	26.2 %	100.0 %
	% within fringe	100.0 %	100.0 %	100.0 %	100.0 %

Notes
(i) Figures in bold are absolute number of sampled respondents in the particular distance zones
(ii) Figures underlined are percentages to the total from that particular distance zone
(iii) Figures below the underlined figures are percentages to the total of that particular variable
Source Based on field survey(2013)

Table 6.12 Aligarh urban fringe: Source of drinking water among sampled households

Zone		Drinking water			Total
		Govt. hand pump	Own hand pump	Submersible	
Immediate urban fringe	Count	34	190	169	393
	% within zone	8.7 %	48.3 %	43.0 %	100.0 %
	% within fringe	26.8 %	46.1 %	77.5 %	51.9 %
Intermediate urban fringe	Count	49	110	39	198
	% within zone	24.7 %	55.6 %	19.7 %	100.0 %
	% within fringe	38.6 %	26.7 %	17.9 %	26.2 %
Distant urban fringe	Count	44	112	10	166
	% within zone	26.5 %	67.5 %	6.0 %	100.0 %
	% within fringe	34.6 %	27.2 %	4.6 %	21.9 %
Total	Count	127	412	218	757
	% within zone	16.8 %	54.4 %	28.8 %	100.0 %
	% within fringe	100.0 %	100.0 %	100.0 %	100.0 %

Notes
(i) Figures in bold are absolute number of sampled respondents in the particular distance zones
(ii) Figures underlined are percentages to the total from that particular distance zone
(iii) Figures below the underlined figures are percentages to the total of that particular variable
Source Based on field survey (2013)

precedes a change in the site. This is evident from the improvements in the infrastructural facilities, amenities and their access in the urban fringe. Better access to a location strengthens the external relations of that place leading to changes in its site. The transformation in availability and access to infrastructural facilities and amenities in the urban fringe has greatly improved. This is also an outcome of change in economic activities as the development of secondary and tertiary activities has helped in developing the infrastructural facilities. Moreover, there is also an improvement in general awareness for accessing and using these available facilities. The urban fringe observes developments in infrastructural facilities in many ways. Some of them are discussed below:

The debilitating effects of 'distance from town' need to be countered by connecting outlying villages with more and better social infrastructures (Krishna and Bajpai 2011). The effects of an advancing technology is to increase the ease by which spatial distance barriers may be overcome by more sophisticated forms of transport and the construction of an efficient transport infrastructure (Lewis and Maund 1976). In addition to socioeconomic factors, decreases in commuting costs due to infrastructure investment are another underlying force in the sprawling expansion of cities (Brueckner 2000). Alonso (1964) cites improvements in transportation infrastructure as one of the primary reasons for a city expanding outwards. Daniels in 1999 supports this idea, noting that new road construction will provide more access to the fringe (ibid.). Heimlich and Anderson (2001) state that infrastructure drives the growth of cities by providing the essential framework for

Table 6.13 Aligarh urban fringe: Household quality of life index (HQLI) for the sampled villages

Immediate urban fringe				Intermediate urban fringe				Distant urban fringe			
Sampled villages	HCI	API	HQLI	Sampled villages	HCI	API	HQLI	Sampled villages	HCI	API	HQLI
Maheshpur	1.84	4.84	6.68	Barautha	0.66	1.11	1.76	Maimari	1.37	3.00	4.37
Barhati	2.49	4.67	7.16	Jarauth	2.54	3.71	6.26	Arrana	0.67	2.05	2.72
Wajidpur Nada	2.64	5.37	8.01	Kothia	2.54	3.71	6.26	Chuharpur	2.48	2.65	5.13
Kwarasi	3.80	7.04	10.84	Nagaula	1.19	2.04	3.23	Ahamadpuri	1.11	1.77	2.88
Talaspur Kalan	3.47	7.27	10.74	Bhankri Khas	1.91	5.43	7.34	Tejpur	0.67	2.05	2.72
Ramgarh Panjoopur	2.96	5.62	8.59	Digsi	2.68	3.00	5.68	Sisroi Nagla Dhak	0.32	0.16	0.48
Dhanipur	3.33	3.96	7.29	Imlani	2.68	3.00	5.68	Peepali	1.50	5.12	6.62
Kasba Kol	3.47	7.27	10.74	Karsua	1.81	3.57	5.38	Daud Garhi	0.32	0.16	0.48
Alinagar	2.96	5.62	8.59	Kishan Garhi	2.68	3.00	5.68	Navalpur	1.11	1.77	2.88
Sarsaul	3.33	3.96	7.29	Jatpura	2.68	3.00	5.68	Gada Khera	0.67	2.05	2.72
Haivatpur Fagoi	2.96	5.62	8.59	Alipur	1.81	3.57	5.38	Sikargarhi	0.32	0.16	0.48
				Rafipur Siya	2.68	3.00	5.68	Jagatpur	1.11	1.77	2.88
				Badwamni	2.68	3.00	5.68	Pratappur	0.67	2.05	2.72
				Bhartua	1.81	3.57	5.38	Jangal Garhi	0.32	0.16	0.48
				Madanpur Chhabila	1.11	1.77	2.88	Padki	1.11	1.77	2.88
								Daupur	0.67	2.05	2.72
								Bhataula	0.32	0.16	0.48
								Parauri	0.67	2.05	2.72

Source Based on field survey (2013)

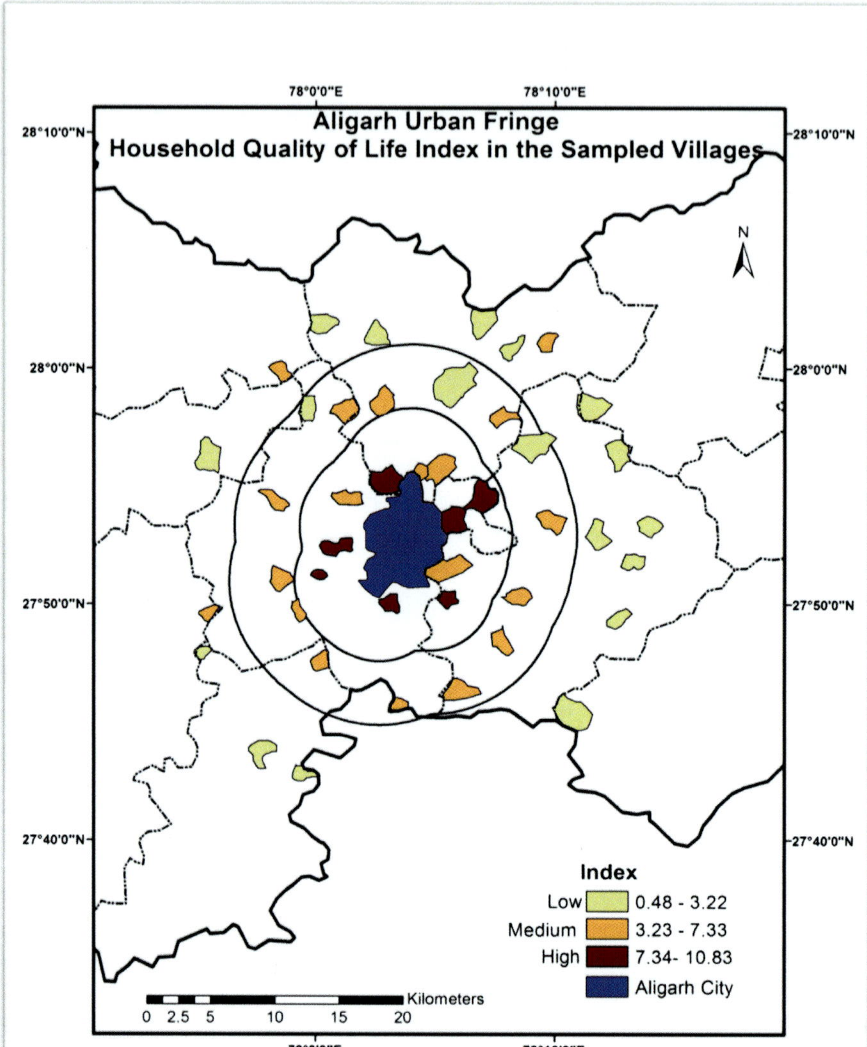

Fig. 6.2 Aligarh urban fringe: Household quality of life index. *Source* Based on field survey(2013)

development (ibid.). Widespread access provided by improvements in transportation infrastructure allows developers to utilize cheap land located outside the city centre (Gillham 2002). In a study published in 2003, Carruthers found that infrastructure investments had mixed effects on growth at the urban fringe.

Urban infrastructure can be defined as the stock of basic facilities, capital equipment and networks that support city functions. Thus water and sewerage systems, roads and rail networks, communications systems, human services facilities such as schools and hospitals are included, along with parks and open space (Wilmoth 2005).

6.3.1 *Improved Accessibility*

The urban fringe of an urban centre forms a pattern depending upon the physiographic and transportation facilities of the area (Sinha 1980). The survey shows that among the total sampled households, members of 526 (70 %) households daily visit to the city for different purposes among which education, marketing and employment are most important one for which the fringe people have to visit daily to the city (Tables 6.14 and 6.15). Beside these the disposal of products like perishable crops and dairy product is another remarkable cause for the daily visit to the city. Distance play an important role for daily visit to the city and this fact is evident from the survey data that from 82 % of the total sampled households in immediate urban fringe daily visit the city while in distant urban fringe its share is only 42 %. This suggests that daily visit to the city depends on the distance from the city.

Rural–urban fringe is characterized by incomplete network of utility services, an inadequate network of public transport a relatively high car ownership ratio and that the majority of workplace and retail purchases of residents of the fringe will be in the urban place itself (Pryor 1969). But all these happenings are not to be found in Indian case excluding planned urban fringe and also not valid for the Aligarh's

Table 6.14 Aligarh urban fringe: Frequency of city visit among sampled households

Zone		Frequency of city visit					Total
		Daily	Weekly	Fortnightly	Monthly	Yearly	
Immediate urban fringe	Count	**323**	**45**	**0**	**15**	**10**	**393**
	% within zone	82.2 %	11.5 %	0.0 %	3.8 %	2.5 %	100.0 %
	% within fringe	61.4 %	42.9 %	0.0 %	23.4 %	22.2 %	51.9 %
Intermediate urban fringe	Count	**133**	**33**	**10**	**13**	**9**	**198**
	% within zone	67.2 %	16.7 %	5.1 %	6.6 %	4.5 %	100.0 %
	% within fringe	25.3 %	31.4 %	58.8 %	20.3 %	20.0 %	26.2 %
Distant urban fringe	Count	**70**	**27**	**7**	**36**	**26**	**166**
	% within zone	42.2 %	16.3 %	4.2 %	21.7 %	15.7 %	100.0 %
	% within fringe	13.3 %	25.7 %	41.2 %	56.3 %	57.8 %	21.9 %
Total	Count	**526**	**105**	**17**	**64**	**45**	**757**
	% within zone	69.5 %	13.9 %	2.2 %	8.5 %	5.9 %	100.0 %
	% within fringe	100.0 %	100.0 %	100.0 %	100.0 %	100.0 %	100.0 %

Notes
(i) Figures in bold are absolute number of sampled respondents in the particular distance zones
(ii) Figures underlined are percentages to the total from that particular distance zone
(iii) Figures below the underlined figures are percentages to the total of that particular variable
Source Based on field survey(2013)

Table 6.15 Aligarh urban fringe: Purpose of city visit among sampled households

Zone		Purpose of daily visitor			Total
		Employment	Education	Marketing	
Immediate urban fringe	Count	**40**	**126**	**157**	**323**
	% within zone	12.4 %	39.0 %	48.6 %	100.0 %
	% within fringe	26.7 %	72.8 %	77.3 %	61.4 %
Intermediate urban fringe	Count	**58**	**37**	**38**	**133**
	% within zone	43.6 %	27.8 %	28.6 %	100.0 %
	% within fringe	38.7 %	21.4 %	18.7 %	25.3 %
Distant urban fringe	Count	**52**	**10**	**8**	**70**
	% within zone	74.3 %	14.3 %	11.4 %	100.0 %
	% within fringe	34.7 %	5.8 %	3.9 %	13.3 %
Total	Count	**150**	**173**	**203**	**526**
	% within zone	28.5 %	32.9 %	38.6 %	100.0 %
	% within fringe	100.0 %	100.0 %	100.0 %	100.0 %

Notes
(i) Figures in bold are absolute number of sampled respondents in the particular distance zones
(ii) Figures underlined are percentages to the total from that particular distance zone
(iii) Figures below the underlined figures are percentages to the total of that particular variable
Source Based on field survey(2013)

urban fringe. High car ratio could not be found in the urban fringe of Aligarh city. The survey shows that only 1.6 % households out of the total sampled households possess cars and use it as means of transport to visit the city (Table 6.17). In India the percentage of commuters from rural–urban fringe to the city for work, recreation and medical services is appreciably higher than for retail purchases. But in the urban fringe of Aligarh city out of the total daily commuters 39 % households reported daily visit to the city for purchasing purpose and its share is higher from the immediate urban fringe than the distant urban fringe (Table 6.15). This is because easy accessibility from the immediate villages makes the households able to visit the city for purchasing purposes and just reverse is the case for distant urban fringe.

6.3.2 *Improved Road Facilities*

Modern age is one of rapid growth of urbanization consequent upon transport development (Sinha 1980). Roadway investments appeared to have no impact on growth in suburban counties, while per capita spending on sewerage products occasionally led to greater growth at the urban fringe (Carruthers 2003). Aligarh city and its urban fringe villages have improved accessibility especially in terms of road transport.

A lot of improvements have been achieved in the last decade with enhanced investments for relaying rural roads and state highways. There has been allocation of Rs.78 crores for maintenance of state highways and Rs. 30 crores for rural roads in the state budget. The urban fringe roads have also benefited from other development programmes (Panchayati Raj, Rural Development and Community Development) which have helped in better accessibility to the city. The improvements in roads have significantly helped the development as well as expansion of urban fringe. In the fringe the length of the total pucca road per lakh population has increased from 199.4 km in 1995 to 355 km in 2012 and the length of the total pucca road per thousand km has increased from 1163.5 km in 1995 to 3036.5 km in 2012. Table 6.16 shows that the villages in immediate urban fringe with pucca road facilities have improved considerably.

But this improvement was not so significant in distant urban fringe. All most all the roads up to 10 km from city boundary were pucca/metalled roads whereas still 11 % of the roads were not metalled in distant urban fringe. This is also evident from the mobility of fringe population due to easy accessibility, more than three-fourth of the sampled households from immediate urban fringe use own means of transport like cycle, bike and car for commuting the city for different purposes (Table 6.17).

Table 6.16 Aligarh urban fringe: Sampled villages with Pucca roads

Zone		Pucca road		Total
		1995	2012	
Immediate urban fringe	Count	**7**	**11**	**11**
	% within zone	63.6 %	100.0 %	100.0 %
	% within fringe	36.8 %	26.2 %	25.0 %
Intermediate urban fringe	Count	**6**	**15**	**15**
	% within zone	40.0 %	100.0 %	100.0 %
	% within fringe	31.6 %	35.7 %	34.1 %
Distant urban fringe	Count	**6**	**16**	**18**
	% within zone	33.3 %	88.9 %	100.0 %
	% within fringe	31.6 %	38.1 %	40.9 %
Total	Count	**19**	**42**	**44**
	% within zone	43.2 %	95.5 %	100.0 %
	% within fringe	100.0 %	100.0 %	100.0 %

Notes
(i) Figures in bold are absolute number of sampled respondents in the particular distance zones
(ii) Figures underlined are percentages to the total from that particular distance zone
(iii) Figures below the underlined figures are percentages to the total of that particular variable
Source Economics and Statistics Division, Sankhyikiya Patrika (1995, 2012)

Table 6.17 Aligarh urban fringe: Means of transport for city visit among sampled households

Zone		Means of transport for city visit						Total
		Cycle	Bike	Tempo	Bus	Car	Other	
Immediate urban fringe	Count	**115**	**202**	**49**	**0**	**9**	**18**	**393**
	% within zone	29.3 %	51.4 %	12.5 %	0.0 %	2.3 %	4.6 %	100.0 %
	% within fringe	81.0 %	78.9 %	34.5 %	0.0 %	75.0 %	78.3 %	51.9 %
Intermediate urban fringe	Count	**25**	**32**	**64**	**71**	**3**	**3**	**198**
	% within zone	12.6 %	16.2 %	32.3 %	35.9 %	1.5 %	1.5 %	100.0 %
	% within fringe	17.6 %	12.5 %	45.1 %	39.0 %	25.0 %	13.0 %	26.2 %
Distant urban fringe	Count	**2**	**22**	**29**	**111**	**0**	**2**	**166**
	% within zone	1.2 %	13.3 %	17.5 %	66.9 %	0.0 %	1.2 %	100.0 %
	% within fringe	1.4 %	8.6 %	20.4 %	61.0 %	0.0 %	8.7 %	21.9 %
Total	Count	**142**	**256**	**142**	**182**	**12**	**23**	**757**
	% within zone	18.8 %	33.8 %	18.8 %	24.0 %	1.6 %	3.0 %	100.0 %
	% within fringe	100.0 %	100.0 %	100.0 %	100.0 %	100.0 %	100.0 %	100.0 %

Notes
(i) Figures in bold are absolute number of sampled respondents in the particular distance zones
(ii) Figures underlined are percentages to the total from that particular distance zone
(iii) Figures below the underlined figures are percentages to the total of that particular variable
Source Based on field survey (2013)

6.3.3 Access and Awareness for Better Education

One of the most important and positive impacts of Aligarh city on urban fringe households have been the awareness to education. The proximity to the city and access to better and specialized education facilities have also helped in overall improvements among sampled households in urban fringe. The sampled households were found to be increasingly concerned for their children's education.

The urban economic environment has also contributed in enhancement of awareness to education. They consider better future of their children by educating them. Significant investments by government (Rs. 53.5 crores for school education and 23.7 crores for education for all programmes) have also improved the availability and access to basic school education. 94 % of total households reported of sending their children to schools. More importantly, the number of dropouts up to primary level was much lesser in urban fringe compared to average rural dropouts in the Aligarh district. The urban fringe households also reported of students daily commuting to Aligarh city for higher education.

In the fringe areas the number of primary schools per lakh population has increased to 75.97 in 2012 from 58.36 in 1995. The number of senior primary schools per lakh population has increased to 34.63 in 2012 from 16.13 in 1995. Along with the increase of primary and senior primary school in the fringe areas, the number of higher secondary schools per lakh population has also increased to 8.33 in 2012 from 3.8 in 1995. The preferential location of high schools and colleges in towns and villages closer to towns make the villagers able to acquire higher education. There is increase in the number of villages having schools from

Table 6.18 Aligarh urban fringe: Sampled villages with primary school

Zone		Primary school		Total
		1995	2012	
Immediate urban fringe	Count	**9**	**11**	**11**
	% within zone	81.8 %	100.0 %	100.0 %
	% within fringe	23.7 %	25.0 %	25.0 %
Intermediate urban fringe	Count	**13**	**15**	**15**
	% within zone	86.7 %	100.0 %	100.0 %
	% within fringe	34.2 %	34.1 %	34.1 %
Distant urban fringe	Count	**16**	**18**	**18**
	% within zone	88.9 %	100.0 %	100.0 %
	% within fringe	42.1 %	40.9 %	40.9 %
Total	Count	**38**	**44**	**44**
	% within zone	86.4 %	100.0 %	100.0 %
	% within fringe	100.0 %	100.0 %	100.0 %

Notes
(i) Figures in bold are absolute number of sampled respondents in the particular distance zones
(ii) Figures underlined are percentages to the total from that particular distance zone
(iii) Figures below the underlined figures are percentages to the total of that particular variable
Source Economics and Statistics Division, Sankhyikiya Patrika (1995, 2012)

Table 6.19 Aligarh urban fringe: Sampled villages with junior high school

Zone		Junior high school		Total
		1995	2012	
Immediate urban fringe	Count	**2**	**5**	**11**
	% within zone	18.2 %	45.5 %	100.0 %
	% within fringe	40.0 %	23.8 %	25.0 %
Intermediate urban fringe	Count	**1**	**9**	**15**
	% within zone	6.7 %	60.0 %	100.0 %
	% within fringe	20.0 %	42.9 %	34.1 %
Distant urban fringe	Count	**2**	**7**	**18**
	% within zone	11.1 %	38.9 %	100.0 %
	% within fringe	40.0 %	33.3 %	40.9 %
Total	Count	**5**	**21**	**44**
	% within zone	11.4 %	47.7 %	100.0 %
	% within fringe	100.0 %	100.0 %	100.0 %

Notes
(i) Figures in bold are absolute number of sampled respondents in the particular distance zones
(ii) Figures underlined are percentages to the total from that particular distance zone
(iii) Figures below the underlined figures are percentages to the total of that particular variable
Source Economics and Statistics Division, Sankhyikiya Patrika (1995, 2012)

Table 6.20 Aligarh urban fringe: Sampled villages with higher secondary school

Zone		Higher secondary		Total
		1995	2012	
Immediate urban fringe	Count	**0**	**4**	**11**
	% within zone	0.0 %	36.4 %	100.0 %
	% within fringe	0.0 %	57.1 %	25.0 %
Intermediate urban fringe	Count	**1**	**1**	**15**
	% within zone	6.7 %	6.7 %	100.0 %
	% within fringe	50.0 %	14.3 %	34.1 %
Distant urban fringe	Count	**1**	**2**	**18**
	% within zone	5.6 %	11.1 %	100.0 %
	% within fringe	50.0 %	28.6 %	40.9 %
Total	Count	**2**	**7**	**44**
	% within zone	4.5 %	15.9 %	100.0 %
	% within fringe	100.0 %	100.0 %	100.0 %

Notes
(i) Figures in bold are absolute number of sampled respondents in the particular distance zones
(ii) Figures underlined are percentages to the total from that particular distance zone
(iii) Figures below the underlined figures are percentages to the total of that particular variable
Source Economics and Statistics Division, Sankhyikiya Patrika (1995, 2012)

1995 to 2012 (Tables 6.18, 6.19 and 6.20). Although in case of primary school distance does not make any difference in the urban fringe of Aligarh city but in case of junior high school, senior high school distance plays an important role.

Significantly, there is overall betterment in the educational facilities and its accessibility. However, again immediate urban fringe villages reported higher share (Tables 6.18, 6.19 and 6.20).

Aligarh Urban Fringe: Many private and convent schools have come up in the fringe area

Aligarh Urban Fringe: Awareness and access to education has improved

Government schools are short of teacher

Convent education is becoming popular in rural areas

Education among girls has improved

6.3.4 Improved Health Facilities

Another positive impact of Aligarh city on its urban fringe villages was observed in awareness and access to health facilities. Increasing number of urban fringe households now approach the modern urban-based health facilities which are not available proper the fringe villages. Qualified doctors are reluctant to have away from the town centre. The fringe dwellers can go to the area for medical services (Sinha 1980). They also had more faith in their traditional nature-based treatment for health problems. Traditionally, the rural population due to ignorance and difficulties in accessing medical facilities depended largely on untrained quacks. Again substantial investments by government (Rs. 30.4 crores for health and 47.9 crores for women and child care) have improved not only health facilities in the city but also in urban fringe's villages and rural health centres.

The sampled urban fringe households have also reported of availing health facilities in the city. This improvement in health facilities along with improved accessibility and access has resulted in escalating number of urban fringe households reporting availing of health facility. Although there is no growth in the number of mother–child centres and family welfare centres immediate zone due to better accessibility avail these facilities more than the distant zone (Tables 6.21 and 6.22).

Table 6.21 Aligarh urban fringe: Sampled villages with mother–child centre

Zone		Mother–child centre		Total
		1995	2012	
Immediate urban fringe	Count	**6**	**6**	**11**
	% within zone	54.5 %	54.5 %	100.0 %
	% within fringe	60.0 %	66.7 %	25.0 %
Intermediate urban fringe	Count	**2**	**1**	**15**
	% within zone	13.3 %	6.7 %	100.0 %
	% within fringe	20.0 %	11.1 %	34.1 %
Distant urban fringe	Count	**2**	**2**	**18**
	% within zone	11.1 %	11.1 %	100.0 %
	% within fringe	20.0 %	22.2 %	40.9 %
Total	Count	**10**	**9**	**44**
	% within zone	22.7 %	20.5 %	100.0 %
	% within fringe	100.0 %	100.0 %	100.0 %

Notes
(i) Figures in bold are absolute number of sampled respondents in the particular distance zones
(ii) Figures underlined are percentages to the total from that particular distance zone
(iii) Figures below the underlined figures are percentages to the total of that particular variable
Source Economics and Statistics Division, Sankhyikiya Patrika (1995, 2012)

Table 6.22 Aligarh urban fringe: Sampled villages with family welfare centre

Zone		Family Welfare Centre		Total
		1995	2012	
Immediate urban fringe	Count	**6**	**6**	**11**
	% within zone	54.5 %	54.5 %	100.0 %
	% within fringe	60.0 %	66.7 %	25.0 %
Intermediate urban fringe	Count	**2**	**1**	**15**
	% within zone	13.3 %	6.7 %	100.0 %
	% within fringe	20.0 %	11.1 %	34.1 %
Distant urban fringe	Count	**2**	**2**	**18**
	% within zone	11.1 %	11.1 %	100.0 %
	% within fringe	20.0 %	22.2 %	40.9 %
Total	Count	**10**	**9**	**44**
	% within zone	22.7 %	20.5 %	100.0 %
	% within fringe	100.0 %	100.0 %	100.0 %

Notes
(i) Figures in bold are absolute number of sampled respondents in the particular distance zones
(ii) Figures underlined are percentages to the total from that particular distance zone
(iii) Figures below the underlined figures are percentages to the total of that particular variable
Source Economics and Statistics Division, Sankhyikiya Patrika (1995, 2012)

6.3.5 Improved Marketing Facilities

The existing city commercial centre, markets have also been developed and improved in the Aligarh urban fringe. There was significant increase of fair price shops and local markets in the study area (Tables 6.23 and 6.24). While the wholesale market remained stagnant but there too volume of trade has increased significantly. The survey again reveals that these market facilities have greater concentration in immediate urban fringe (Table 6.25).

6.3.6 Improved Electricity Supply

All the sampled villages of Aligarh urban fringe were made electrification and 79 % of the sampled households have also electric connection in their houses. Distance from the city also makes difference in the connection of electricity in the households. While in the immediate urban fringe 93 % households have electricity connection, in the distant urban fringe only 57 % households had the connection despite all the villages in this zone were electrified (Table 6.26). However, the electricity supply is greatly varied. The city has an average daily supply of 16 h while rural areas have 8–12 h supply. The government has allocated Rs. 83 crores for rural electrification but respondents were not satisfied with the supply of electricity. Interestingly, all urban fringe respondents acknowledged the city proximity

Table 6.23 Aligarh urban fringe: Distribution of fair price shop

Zone		Fair price shop		Total
		1995	2012	
Immediate urban fringe	Count	**9**	**11**	**11**
	% within zone	81.8 %	100.0 %	100.0 %
	% within fringe	25.7 %	27.5 %	25.0 %
Intermediate urban fringe	Count	**12**	**14**	**15**
	% within zone	80.0 %	93.3 %	100.0 %
	% within fringe	34.3 %	35.0 %	34.1 %
Distant urban fringe	Count	**14**	**15**	**18**
	% within zone	77.8 %	83.3 %	100.0 %
	% within fringe	40.0 %	37.5 %	40.9 %
Total	Count	**35**	**40**	**44**
	% within zone	79.5 %	90.9 %	100.0 %
	% within fringe	100.0 %	100.0 %	100.0 %

Notes
(i) Figures in bold are absolute number of sampled respondents in the particular distance zones
(ii) Figures underlined are percentages to the total from that particular distance zone
(iii) Figures below the underlined figures are percentages to the total of that particular variable
Source Economics and Statistics Division, Sankhyikiya Patrika (1995, 2012)

Table 6.24 Aligarh urban fringe: Distribution of local market

Zone		Local market		Total
		1995	2012	
Immediate urban fringe	Count	**3**	**4**	**11**
	% within zone	27.3 %	36.4 %	100.0 %
	% within fringe	100.0 %	80.0 %	25.0 %
Intermediate urban fringe	Count	**0**	**0**	**15**
	% within zone	0.0 %	0.0 %	100.0 %
	% within fringe	0.0 %	0.0 %	34.1 %
Distant urban fringe	Count	**0**	**1**	**18**
	% within zone	0.0 %	5.6 %	100.0 %
	% within fringe	0.0 %	20.0 %	40.9 %
Total	Count	**3**	**5**	**44**
	% within zone	6.8 %	11.4 %	100.0 %
	% within fringe	100.0 %	100.0 %	100.0 %

Notes
(i) Figures in bold are absolute number of sampled respondents in the particular distance zones
(ii) Figures underlined are percentages to the total from that particular distance zone
(iii) Figures below the underlined figures are percentages to the total of that particular variable
Source Economics and Statistics Division, Sankhyikiya Patrika (1995, 2012)

Table 6.25 Aligarh urban fringe: Distribution of wholesale market

Zone		Wholesale market		Total
		1995	2012	
Immediate urban fringe	Count	**1**	**1**	**11**
	% within zone	9.1 %	9.1 %	100.0 %
	% within fringe	100.0 %	100.0 %	25.0 %
Intermediate urban fringe	Count	**0**	**0**	**15**
	% within zone	0.0 %	0.0 %	100.0 %
	% within fringe	0.0 %	0.0 %	34.1 %
Distant urban fringe	Count	**0**	**0**	**18**
	% within zone	0.0 %	0.0 %	100.0 %
	% within fringe	0.0 %	0.0 %	40.9 %
Total	Count	**1**	**1**	**44**
	% within zone	2.3 %	2.3 %	100.0 %
	% within fringe	100.0 %	100.0 %	100.0 %

Notes
(i) Figures in bold are absolute number of sampled respondents in the particular distance zones
(ii) Figures underlined are percentages to the total from that particular distance zone
(iii) Figures below the underlined figures are percentages to the total of that particular variable
Source Economics and Statistics Division, Sankhyikiya Patrika (1995, 2012)

Table 6.26 Aligarh urban fringe: Electricity supply among the sampled households

Zone		Electricity supply in the households		Total
		Yes	No	
Immediate urban fringe	Count	**364**	**29**	**393**
	% within zone	92.6 %	7.4 %	100.0 %
	% within fringe	60.5 %	18.7 %	51.9 %
Intermediate urban fringe	Count	**143**	**55**	**198**
	% within zone	72.2 %	27.8 %	100.0 %
	% within fringe	23.8 %	35.5 %	26.2 %
Distant urban fringe	Count	**95**	**71**	**166**
	% within zone	57.2 %	42.8 %	100.0 %
	% within fringe	15.8 %	45.8 %	21.9 %
Total	Count	**602**	**155**	**757**
	% within zone	79.5 %	20.5 %	100.0 %
	% within fringe	100.0 %	100.0 %	100.0 %

Notes
(i) Figures in bold are absolute number of sampled respondents in the particular distance zones
(ii) Figures underlined are percentages to the total from that particular distance zone
(iii) Figures below the underlined figures are percentages to the total of that particular variable
Source Based on field survey(2013)

as advantageous for better infrastructure and amenities but they blame city for poor electric supply as they feel that their share of supply is being diverted for city residents.

Table 6.27 Aligarh urban fringe: Infrastructural facility index (IFI) for the sampled villages

Immediate urban fringe						Intermediate urban fringe						Distant urban fringe					
Sampled villages	TFI	EFI	HFI	MFI	IFI	Sampled villages	TFI	EFI	HFI	MFI	IFI	Sampled villages	TFI	EFI	HFI	MFI	IFI
Maheshpur	6.15	8.4	11.8	0.91	27.26	Barautha	20.25	10.2	24.8	0.91	56.16	Maimari	6.15	13.4	0	0.91	20.46
Barhati	6.15	5.2	11.4	9.81	32.56	Jarauth	12.95	10.2	0	0.91	24.06	Arrana	14.25	10.2	0	0.91	25.36
Wajidpur Nada	7.75	5	10.2	9.81	32.76	Kothia	0.45	5	0	0.91	6.36	Chuharpur	7.25	10.2	16	0.91	34.36
Kwarasi	20.25	13.4	96.9	9.81	140.36	Nagaula	6.15	10.2	0	0.91	17.26	Ahamadpuri	12.95	10.2	0	0.91	24.06
Talaspur Kalan	12.95	5	7.7	9.81	35.46	Bhankri Khas	14.55	5.2	24.8	0.91	45.46	Tejpur	6.15	5.2	0	0.91	12.26
Ramgarh Panjoopur	6.15	10.2	32.2	0.91	49.46	Digsi	20.25	5	24	0.91	50.16	Sisroi Nagla Dhak	7.25	0	0	0.91	8.16
Dhanipur	7.75	18.6	87	10.61	123.96	Imlani	2.25	5	0	0.91	8.16	Peepali	7.75	0	0	0	7.75
Kasba Kol	7.75	18.6	52.6	9.81	88.76	Karsua	0.45	5.2	0	0.91	6.56	Daud Garhi	12.95	0	0	0.91	13.86
Alinagar	12.95	10.2	3.6	9.81	36.56	Kishan Garhi	0.45	13.4	1.4	0.91	16.16	Navalpur	6.15	0	0	0	6.15
Sarsaul	20.25	13.4	13.4	10.61	57.66	Jatpura	6.15	10.2	7.5	0.91	24.76	Gada Khera	5.7	5	0	0.91	11.61
Haivatpur Fagoi	7.75	5	8	9.81	30.56	Alipur	0.45	13.4	7.5	0.91	22.26	Sikargarhi	0.45	0	0	0.91	1.36
					56.16	Rafipur Siya	0.45	0	0	0.91	1.36	Jagatpur	13.45	10.2	0	0.91	24.56
					24.06	Badwamni	0.45	5.2	0	0.91	6.56	Pratappur	0.45	0	0	0.91	1.36
					6.36	Bhartua	0.45	10.2	3.7	0.91	15.26	Jangal Garhi	7.25	0	16	0.91	24.16
					17.26	Madanpur Chhabila	6.15	0	0	0	6.15	Padki	14.1	0	8	0.91	23.01
												Daupur	7.75	0	0	0	7.75
												Bhataula	6.15	13.6	0	0.91	20.66
												Parauri	6.15	10.2	0	9.81	26.16

Source Economics and Statistics Division, Sankhyikiya Patrika (2012)

6.3.7 *Infrastructural Facility Index*

To assess the status of infrastructural facilities for sampled villages in the urban fringe of Aligarh city, an Infrastructural Facility Index was calculated on the basis of 11 variables. Infrastructural Facility Index was made up of four distinctive indices such as transport facility index (TFI), education facility index (EFI), health facility

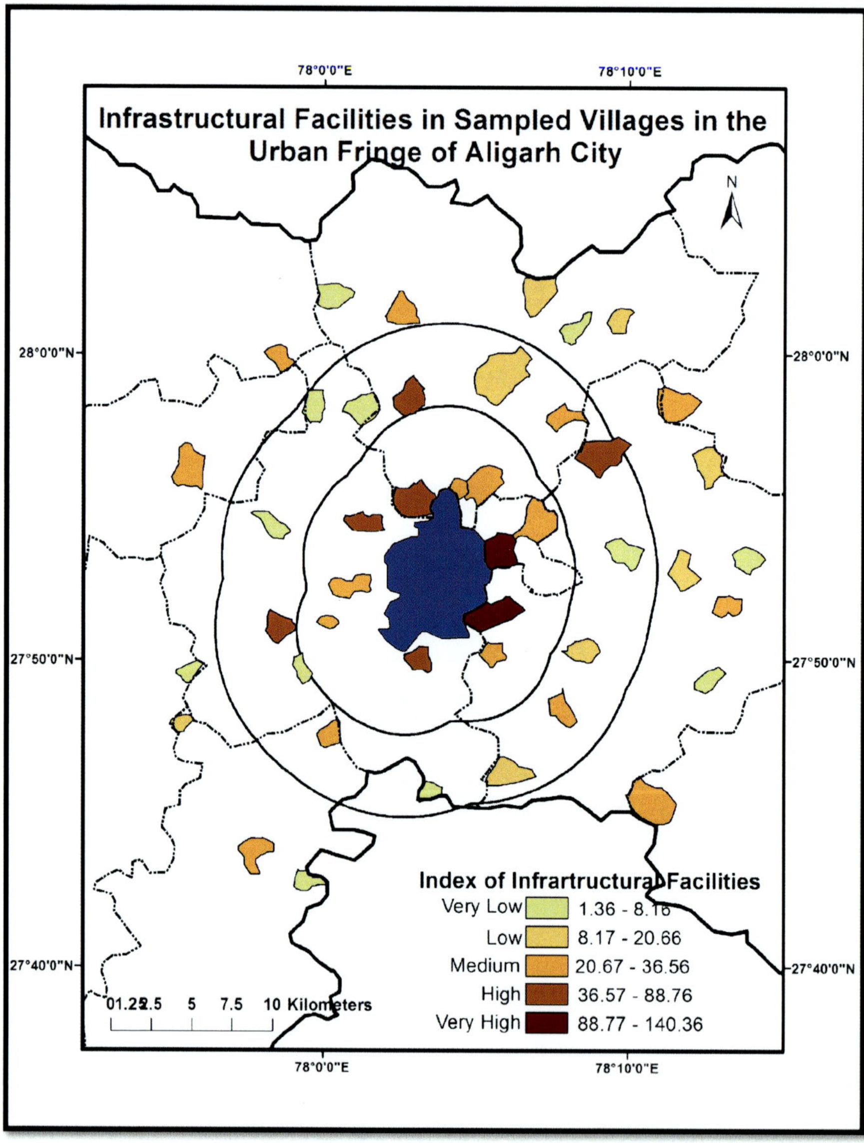

Fig. 6.3 Aligarh urban fringe: Infrastructural facility index. *Source* Based on field survey (2013)

(HFI) and marketing facility index (MFI). The infrastructural facility index is made by proportionate possession weighting technique. All unidirectional variables were assigned a weight. The percentage of villages availing with a particular facility was assigned score out of ten. The weight for the particular facility is calculated by deducting the value of assigned score from ten. Those villages with the greater score on this index would have higher scores. After adding up all the scores of different variables of a particular observation form the index. Here, positive aspects of the infrastructural facilities were selected to make the index hence, higher the score indicates the better infrastructural condition in the villages. Selected 11 variables are under four indices. The details of variables are given below

$$\mathbf{IFI = TFI + EFI + HFI + MFI}$$

(a) *Transport facility index (TFI)*

Sampled villages have following facilities

(i) Pucca road
(ii) Bus station

(b) *Education facility index (EFI)*

Sampled villages have following facilities

(i) Primary school
(ii) Junior high school
(iii) Higher secondary school
(iv) Optional educational centre

(c) *Health facility index (HFI)*

Sampled villages have following facilities

(i) Mother–child centre
(ii) Family welfare centre

(d) *Marketing facility index (MFI)*

Sampled villages have following facilities

(i) Fair price shop
(ii) Retail market
(iii) Wholesale market

The index reveals that only two sampled villages possess very high score of the index and these two belong to the immediate urban fringe (Fig.6.3). These villages enjoy better infrastructural facilities than others. Three of the six villages having high score belong to immediate urban fringe and rest three belong to intermediate urban fringe. Rest of the six villages out of the total sampled villages in immediate urban fringe possess medium score. From here it is clear that all the sampled villages in immediate urban fringe do not suffer from poor infrastructural facilities while in distant urban fringe all the villages possess either low or very low score. This means distance from the city centre has remarkable effects on the infrastructural facilities in the sampled villages in the urban fringe (Table 6.27). Even, Krishna and Bajpai in 2011 also stated that the disadvantages of distance have been compounded by public policies.

References

Alonso W (1964) Location and land use: toward a general theory of land rent. Harvard University Press, Cambridge

Bryant CR (1995) The role of local actors in transforming the urban fringe. J Rural Stud 11:255–267

Brueckner JK (2000) Urban Sprawl: diagnosis and remedies. International Regional Science Review 23(2):160–71

Carruthers JI (2003) Growth at the fringe: the influence of political fragmentation in United States Metropolitan Areas. Papers Reg Sci 82:472–499

Diaskosavvas D (2008) How far can agricultural policies contribute to achieving rural development objectives in semi-urban areas? www.ruralitynearbythecity.com

FAO (2003) Overview of land value conditions, AGL/MISC/35/2003, FAO, Rome ftp://ftp.fao.org/agl/agll/docs/misc35e.pdf

Fazal S (2013) Land use dynamics in a developing economy. Regional perspective from India. Springer Publication, The Netherlands

Gillham O (2002) The limitless city: a primer on the urban sprawl debate. Island Press, Washington DC

Heimlich RE, Anderson WD (2001) Developing the Urban Fringe and Beyond: impacts on agriculture and rural land, Report Number 803. USDA Economic Research Service: Washington, DC. Available online at http://www.ers.usda.gov/media/536851/aer803_1_.pdf

Krishna A, Bajpai D (2011) Linear spread and radial dissipation: experiencing growth in rural India, 1993–2005. Econ Polit Weekly 46(38):44–51

Lewis GJ, Maund DJ (1976) The urbanization of the countryside: a framework for analysis. Geografiska Annaler Ser B Human Geogr 58(1):17–27

Pryor RJ (1969) Delineating outer suburban and urban fringe. Geografiska Annaler Ser B Human Geogr 51:33

Ramachandran R (1989) Urbanisation and urban systems in India. Oxford University Press (Reprint 2001), Delhi

Registrar General 1981, 1991, 2001, 2011, India, 2A. Census of India. Mansingh Road, New Delhi 110011

Sankhyikiya Patrika 1995, 2012 Economics and statistics division. http://updes.up.nic.in/spatrika/spatrika.htm

Saville J (1957) Rural depopulation in England and Wales, 1851–1951. Routledge and Kegan Paul, London

Sinha MMP (1980) The impact of urbanization on land use in the rural-urban fringe: a case study of Patna. Concept Publishing Company, New Delhi

Wilmoth D (2005) Urban infrastructure and Metropolitan planning: connection and disconnection. https://www.griffith.edu.au/__data/assets/pdf_file/0009/81396/infrastructure-16-wilmoth.pdf

Chapter 7
Urban Fringe: Sustainability of Livelihoods and Well-Being of Urban Fringe Households

India is witnessing transformations in economic, social and demographic spheres. These transformations are more pronounced in the past few decades. The country from the regime of centralized economic planning along with extensive regulatory control over the economy has shifted to substantially economically liberalized mode. This shift has its bearing on all sectors. The GDP has grown from 1.1 % in 1991 to 6.9 % in 2014. The share of agriculture in GDP has fallen up to 13.7 %, similarly the share of non-farm income and employment in the total income and employment of rural households have increased. The emergence of non-farm sector has augmented urbanization process resulting in urban expansions.

Over the first half of the twentieth century, growth in Indian cities remained largely confined within municipal boundaries and much of the population growth and migration to cities has been accommodated by crowding more and more people into its existing urban areas rather than by expansion of cities into suburbs and fringe areas (Brush 1968). This pattern of growth and population absorption has, however, changed significantly over the past two to three decades (Shaw 2005). The urban centres in India have grown not only in population but also in their areal spread (Shaw 1999). The urban sprawl affects agricultural land leading to its scarcity (Fischer and Sun 2001; Fazal 2014). This change in turn definitely poses threats for present and future generations. A key challenge to the urbanization process is the rapid conversion of large amount of prime agricultural land to urban land use as well as transformation in the livelihoods of urban fringe dwellers (Owusu and Agyei 2007).

Competition for available land in an open market system leads to an increase in land prices (FAO 2003). This status of land resource is highly dynamic at the urban margins, which is termed as urban fringe. The urban fringe is a social, economic and environmental space where three systems (agricultural, urban and the natural) are in constant interaction (Fazal 2013). A combination of factors is responsible for the emergence of peri-urbanization or urban sprawl into the urban fringe, creating melting transitional urban fringe zones.

N. Banu and S. Fazal, *Livelihood and Wellbeing in the Urban Fringe*,
The Urban Book Series, DOI 10.1007/978-3-319-39660-6_7

Although, the urban fringe, which comprised of well-developed agricultural environs, is under threat of urban expansion resulting in not only loss of fertile agricultural land but also loss of livelihood option (Banu and Fazal 2013). Many scholars have opined that urbanization is also blessing for the urban fringe residents in economic terms. The emergence of urban monetary economy which allows the quantification of every commodity in monetary terms subjects people living in peripheral villages to hardships. From the positive side, urban expansion creates opportunities in wage employment and trading for people in urban fringe area, and provides them with access to services and infrastructure (Aberra and King 2005).

7.1 Sustainability and Urban Fringe

Scholars have defined and interpreted the concept of urban fringe sustainability in different ways. As a place of transforming and growing settlement, where the place needs to be fulfilling the present requirement without compromising the ability of future generations. Here the urban fringe sustainability needs to be understood as livelihood requirements of population dependent on them and should not be the capability/opportunity to use the resources. As the greater capability/opportunity adversely affects the sustainability of urban fringe by the dynamic and changing flows of commodities, capital, natural, social and economic resources (Allen et al. 1999).

7.2 Dualism of Urbanization in Urban Fringe

The urbanization process is quite complex and so is its impact on urban and rural environs. Depending upon the ongoing stage and dynamism, these entities have different level of impacts. It may be negative or positive and sometimes both negative and positive impacts can be observed simultaneously. Urbanization process can never be neutral in the dynamics of rural communities. More over rural transformation does not require specific relationship between rural and urban entities, the progression of urban centre has always its bearing on the environs of urban fringe, however, the magnitude may vary. Each element of the environment (natural, social and economic) is under stress in urban fringe (Lyon 1983). The effects of urbanization also has significant impact on the people of urban fringe, depending on their ability to adopt, the urban fringe may be a place of countless opportunities or may be a vulnerable place for their survival (Abbas et al. 2013).

7.3 Urbanization and Its Effect on Aligarh's Urban Fringe Livelihood

Urbanization affects land changes through the transformation of urban–rural linkages and this process with transformations in economy, and immigration, augments considerable land use change in the recent times. The changes in land use pose a serious threat to urban fringe livelihood because most households in the urban fringe area depends on land either for food, water or fuel wood (Abbas et al. 2013). The study area has one of the favourable physical conditions for agricultural activity in the country as it lies in the fertile agricultural tract between Ganga and Yamuna Rivers. Therefore, any loss to agricultural land here can have far reaching implications. There was a tremendous increase in urban area between 1980 and 2010 (230 %) while a total of 5448 hectares of fertile crop land were lost due to city's expansion (Fazal 2014). In addition, agriculture activities are weakened, as farmers also move to engage themselves in other economic activities which attract immediate livelihood goals as compared to agriculture returns (Banu and Fazal 2014). While the continuing struggle against land acquisition for instance by farmers in Aligarh, which took a violent turn, and became a political ploy is being projected as a battle by farmers for big money, the reality is that a majority of the farmers do not want to dispense with their ancestral land and rather they are being forced to do so (Sharma 2010). So, whatever the cause whether the crop land is shrinking by the willingness of the farmers or by the forces operating on them, the ultimate effects is on their livelihoods.

7.4 Livelihood Assets in Aligarh's Urban Fringe

The dynamic processes of socio-economic and environmental change which are usually a major element of the urban fringe are likely to have an impact on the opportunities and constraints faced by different groups in their access to assets and the construction of livelihood strategies. The five capital assets in the framework are: physical, human, natural, social and financial. Assets are not only resources that people use; they are also what give people the capability to act. For example, human capital gives people the tools to change their capacity to improve their circumstances. As the person's asset increases they are better place to protect themselves from shocks and their vulnerabilities decreases.

An asset hexagon allows us to visualize the variation in each of the five classes of assets simultaneously including Composite Asset Index (Fig. 7.1). The asset-based approach thereby emphatically illustrates the all-around favourable asset base of the Aligarh's urban fringe. The intermediate urban fringe has relatively high scores for social capital as well as for Composite Asset Index. While, the

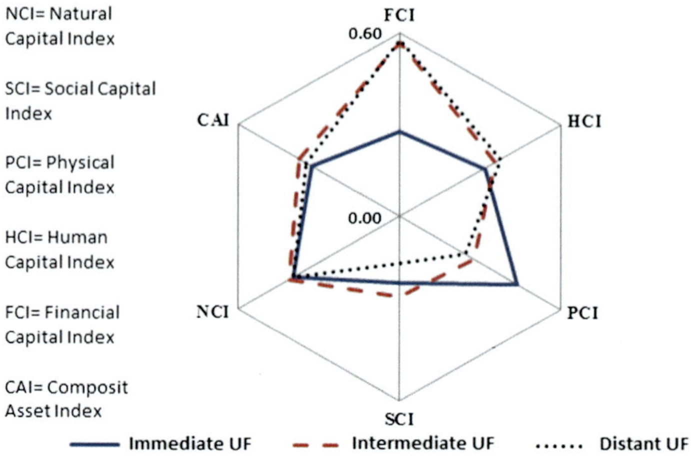

Fig. 7.1 Livelihood asset polygons for the Aligarh's urban fringe. *Source* Field Survey, 2013 and Census of India, 2011

immediate urban fringe has relatively favourable scores for physical capital but it tends to score low in financial, human and social capital which results in low score for composite asset index. The distant urban fringe is overlapping with intermediate urban fringe in the score of financial, human and natural capital index but it has significantly low score in social capital. In Composite Asset Index, Intermediate urban fringe has the highest score whereas immediate urban fringe has the lowest on the Composite Asset Index.

7.5 Coping Strategies for Livelihood Stress

The economy in urban fringe is transforming from predominantly rural agrarian to predominantly urban oriented. With this changing phenomenon, it is inevitable that urban fringe residents need to transform their livelihood especially those who were dependent on natural resource-based activities and they have to adapt and develop a range of strategies by which they can cope with the change. Because of this contextual stresses, livelihood strategies temporarily take the shape of safety mechanisms called 'coping strategies'. These are short-term responses to secure livelihood in periods of stress. With the continuous expanding urban areas, livelihood transformation is inevitable. The present study portrays the livelihood strategies which are adapted by the urban fringe residents and these strategies include diversification, livelihood transformation, crop diversification and migration with the aim of strengthening household resilience by enhancing income (Fig. 7.2). The options open to households in the communities vary according to sources of livelihood and access to livelihood resources.

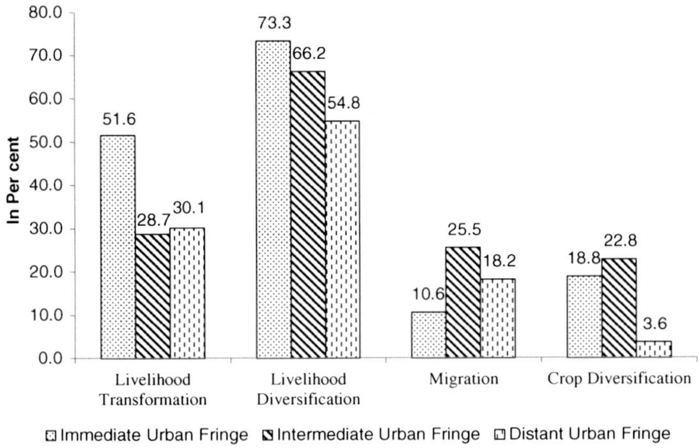

Fig. 7.2 Coping strategies adapted by sampled households in Aligarh's urban fringe. *Source* Field Survey, 2013

The urban fringe is influenced by urban monetized economy, larger market and greater opportunities. The most important element of livelihood strategies in the urban fringe (both in rural and in urban areas) is income diversification. All this induce the urban fringe households to diversify their livelihood for better and improved living. The present study reveals that in Aligarh's urban fringe, income diversification is quite common where increasing number of households is combining non-farming activity with their traditional farming activity. The survey shows that 64.7 % of the total sampled households have adapted diversification as coping strategy and their income coming from more than one economic sector and this is more common in immediate urban fringe and decreases with decreasing accessibility to the city (Fig. 7.2). This is reflected in Physical Asset Index, immediate urban fringe zone has far better score than the intermediate and distant urban fringe (Fig. 7.1) that means infrastructural facilities are better in immediate urban fringe and with increasing distance it decreases (Banu and Fazal 2013). This diversification is necessary for enhancement in earnings required for survival in urban fringe.

Another coping strategy observed in Aligarh's urban fringe is livelihood transformation. The 2011 Population Census of India showed that there is a declining trend in the share of agricultural workers as compared to 1981, 78 % out of the total main workers were associated with agriculture while in 2011, it decreased to 28 %. The reason for this transformation is decreasing land for farming activity, which means small farmers do not have choice but to opt for urban-based livelihood, shedding their traditional farming activity. Again like diversification, households in immediate urban fringe adapted the livelihood transformation as coping strategy more than other two zones because there is more intense pressure on immediate urban fringe than the distanced zones due to the physical expansion of the city just beyond its limit.

In response to the declining agricultural land in the communities, landowners switched to intensification of agriculture as well as diversification of crops. The farmers started cultivating early-maturing and high-yielding crops. The survey results indicate that respondents resort to crop diversification to cope with urbanization. Crop diversification in the study areas includes growing multiple food crops on a field and the choice of crops was guided by the demands in city market (Banu and Fazal 2013). This strategy is adopted to secure livelihood and reduce risk associated with mono-cropping. Here, urbanization acts as a pull factor creating large market for the urban fringe farmers. Farmers in intermediate urban fringe are found to be progressive farmers taking agriculture activity as a business and their livelihood is largely based on agriculture. They take advantage of relative proximity to the city but they also have to absorb the stress of urban shadow effect, which in all sense affects agricultural activity here. They also take risks and grow crops such as vegetables, flowers, chara etc., which they grow mostly for the urban market. Whereas, farms at distant urban fringe have shown subsistence orientation in their cropping pattern, as the cropping pattern was dominated by food grains.

One of the coping strategies observed in the Aligarh urban fringe is to work in the urban area leaving native place for short or long period of time. This migration is mostly male dominated to fulfil income needs. A good number of migrants from urban fringe, especially those who do not have land, sell their labour to non-farm sector in city while their other household members work in farming sector in their native rural areas. This pattern of livelihood diversification observed in the urban fringe is essential because it reduces dependence on one source of livelihood as well as increases the income of household which reduces vulnerability to stresses of livelihood in urban economy. Similarly in Aligarh's urban fringe the household's members are migrating to large urban centres[1] and they are performing better to cope with the stress and sustain their livelihoods rather than the households who adapted other strategies. In Aligarh's urban fringe, migration as coping strategy is adopted more by the intermediate urban fringe than other zone. This results in higher score for social capital as their social networks help them in migrating, settling and securing jobs.

7.6 Measurement of Sustainable Livelihood Index (SLI)

The present study adopts multidimensional, integrated and rational approach targeted to understand the livelihood of households (ranging from rich with greater access to resources or poor with restricted access to resources) (DFID 1999; May et al. 2009). The study evaluates sustainability of livelihood focusing five

[1]Most of the migrants choose their destination to New Delhi which is the capital of India and Faridabad (an industrial town lies under National Capital Region of Delhi). Aligarh is located in the urban shadow of the national capital where they are mostly engaged in industries and factories.

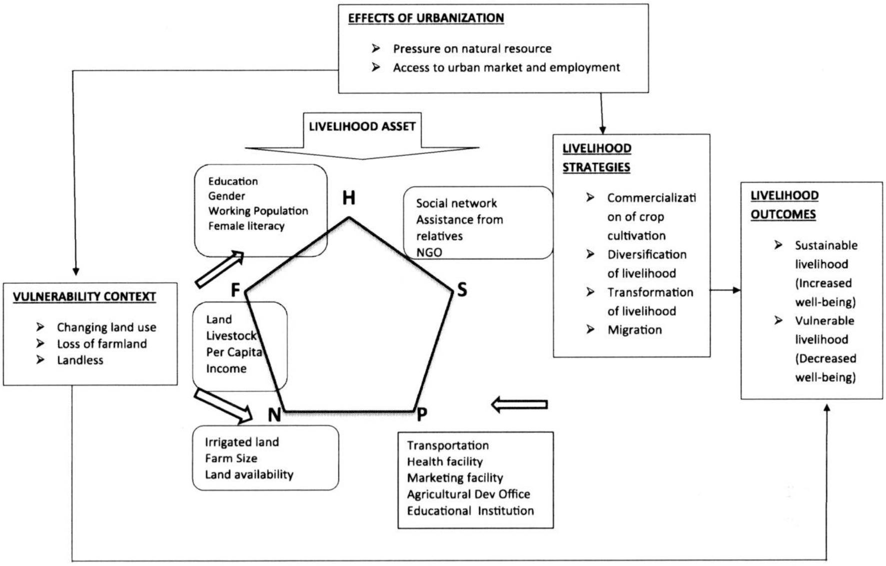

Fig. 7.3 Sustainable urban fringe livelihoods: a framework for analysis

household level assets—natural, social, financial, physical and human capital (Chambers and Conway 1992). The study assumes that process of peri-urbanization adds complexity to household livelihood security and the process of peri-urbanization has been considered as environmental stress (Fig. 7.3).

The present study uses three major-components: Asset Index, Exposure Index and Sensitivity Index for measuring Sustainable Livelihood Index (SLI). Each of these components comprises of several indicators and sub-components. Each sub-component is comprised of several indicators or minor-components and further minor-components of assets are comprised of several indicators (Table 7.1).

The SLI uses a balanced weighted average approach where each sub-component contributes equally to the overall index even though each major-component comprises several sub-components. The SLI formula uses the simple approach of applying equal weights to all major-components, because each of the sub-components is measured on a different scale. Further these were standardized as an index.

$$\mathbf{Index}_{S_d} = \frac{S_d - S_{\min}}{S_{\max} - S_{\min}} \quad (7.1)$$

where, S_d is the original indicator for Aligarh urban fringe villages, and S_{min} and S_{max} are the minimum and maximum values, respectively, for each indicator. After adding up all the scores of different variables of a particular observation, construct the index. In this study, each index is scaled from 0 to 1. In Asset Index and

Table 7.1 Major-components, sub-components and minor-components comprising the sustainable livelihood index (SLI) developed Aligarh's urban fringe

Major-components	Sub-components	Minor-components
Assets	Human capital	Working population, female literacy, education, gender
	Social capital	Social network, assistance from relatives, NGO
	Physical capital	Transport and communication, health facility, education facility, facility for agricultural development, banking facility, marketing facility
	Natural capital	Irrigated land, farm size, land availability
	Financial capital	Land, livestock, per capita income
Exposure	Economic vulnerability	Landless, disposal of land
Sensitivity	Housing condition	Bricked house, sanitation, ventilation, electrification
	Asset possession	Cycle, bike, car, cooler, refrigerator, washing machine, television, submersible

Sensitivity Index, higher the scores indicate better conditions but in Exposure Index, higher scores indicate poor conditions.

The present study, an alternative method for measuring SLI was also calculated which is adapted from the IPCC vulnerability index (IPCC 2001). The understanding is that peri-urbanization is causing vulnerability to livelihood of the population. The livelihood sustainability has three components dependent on assets, exposure and sensitivity. Thus, the measurement of SLI is done by deducting Asset Index from Exposure Index and further multiplying by Sensitivity Index scores. Here, the scale of SLI varies from −1 (least vulnerable or most sustainable) to +1 (highly vulnerable or least sustainable).

$$\mathbf{SLI} = (\mathbf{Exposure\ Index} - \mathbf{Asset\ Index}) \times \mathbf{Sensitivity\ Index} \quad (7.2)$$

7.7 Livelihood Sustainability in Urban Fringe

The closer the location to an urban centre, the more prominent are information and service interactions, the farther the locations the more basic and intermittent the social relationships. Village residents who live closer to towns can more easily benefit from these linear radial patterns of economic development, their lands grow in value and their labour is in greater demand. Those who live further out have comparatively little to offer despite of equal value. Their lands are not attractive to builders and industrialists (Lewis and Maund 1976). But on the other hand, economic development directly affects natural resources and livelihood options in the urban fringe area. People who are not able to improve their livelihoods often fail to

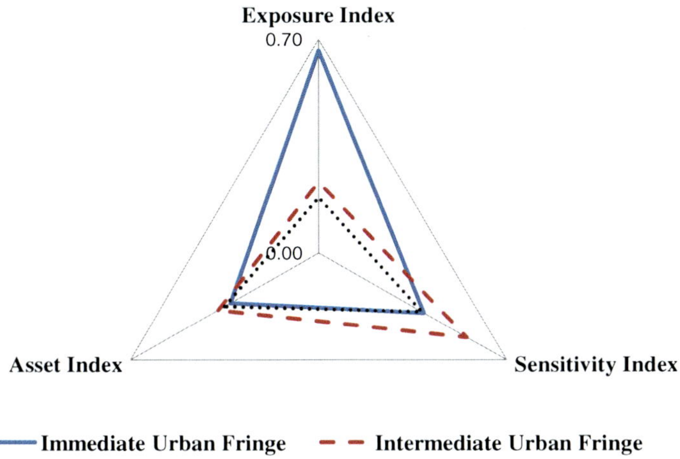

Fig. 7.4 Exposure triangle diagram of the contributing scores for Aligarh's UF. *Source* Field Survey, 2013

access, defend and capitalize on their existing assets and their vulnerability increases (Hahn et al. 2009).

The study reveals that after getting handsome amount by selling their fertile agricultural land, urban fringe residents are unable to utilize this money properly as they are not familiar with monetary investments and dealt only with exchanging land and other material assets. They are found to acquire modern asset from their land money to fulfil their desires. This is common feature in immediate urban fringe. But after losing their asset (fertile agricultural land) they look for alternative means of livelihood which is not very easy for them. Most of the immediate UF people were adopting livelihood diversification and transformation as coping strategy. They are engaged essentially in non-farm activities offered by the city which includes petty trading/business, artisan and constructional labour work which are low income generating livelihood options. The immediate urban fringe households are enjoying the diversified job opportunities created through urbanization as well as the infrastructural facilities but goal to achieve the alternative sustainable livelihood is lacking and that is why although in asset index immediate urban fringe match the score obtained by intermediate urban fringe but in sustainable livelihood framework this zone is more vulnerable than intermediate urban fringe (Fig. 7.4).[2]

As immediate urban fringe is more vulnerable in vulnerability index, for Livelihood Sustainable Index its scores are poor (Fig. 7.5). For this study, score for

[2]The variables of asset, sensitivity and exposure are not unidirectional. The variables of asset and sensitivity indicate positive aspects whereas the variables of vulnerability indicate the negative one. So, higher scores in Asset Index as well as in Sensitivity Index represent better condition but higher scores in Exposure Index stands for worse condition.

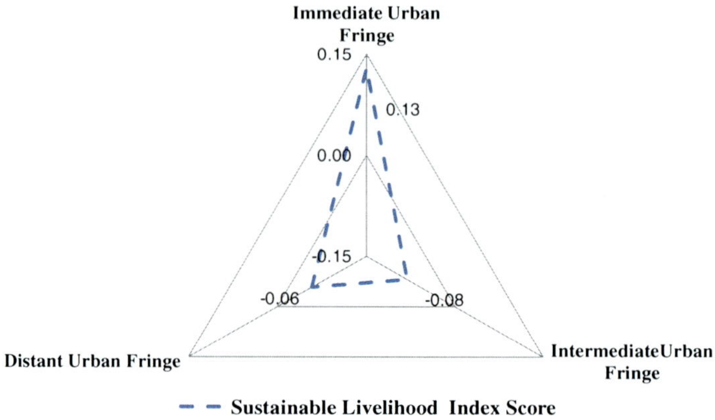

Fig. 7.5 Sustainability triangle diagram of the contributing scores for Aligarh's UF. *Source* Field Survey, 2013

SLI varies from −0.15 to +0.15 and negative value presents better condition while poor condition represented by the positive value. The SLI for immediate urban fringe has the score 0.13 indicating lesser sustainable whereas intermediate urban fringe scores −0.08 which indicates this zone is better in the context of sustainability. Distant urban fringe also presents better condition than immediate urban fringe, which is because of lesser stress of urbanization despite lesser access to opportunities offered by the city.

7.8 Stages of Well-Being

The present study put forward the idea of 'livelihoods ladder' for measurement of degree of poverty and vulnerability depending on their gains and losses of asset. This means that a household add to their existing asset base their position on the ladder moves up, while on the other hand if they lose (because of any reason) then they fall back down the ladder. On the basis of SLI's score the urban fringe villages have been categorized into three stages which include accumulating stage, adapting stage and surviving stage (May et al. 2009). In livelihood ladder most of the villages in immediate urban fringe falls under surviving stage in which life is crawling with sufferings and people in this stage are unable to cope with stress of urbanization. Due to more city proximity to city, this zone has to face the brunt of urban impacts more. All the physical expansion has been occurring in this zone at the cost of very fertile agricultural land on which most of the people's livelihood depended. But after disposal of lands, they adapted alternative livelihoods in non-farm sector.

But due to lower level of educational and technical expertise, they are employed only in lower level, labour-oriented jobs, that too only for male members, thus they

Table 7.2 Stages of Aligarh's urban fringe villages on the livelihoods ladder

Zone	Name	Asset index	Sensitivity index	Exposure index	Sustainable livelihood index	Stages of livelihood ladder
Immediate urban fringe	Maheshpur	0.28	0.33	0.53	0.08	Adapting
	Barhati	0.29	0.41	0.64	0.14	Surviving
	Wajidpur Nada	0.30	0.28	0.57	0.07	Adapting
	Kwarasi	0.41	0.54	0.51	0.05	Adapting
	Talaspur Kalan	0.33	0.39	0.68	0.13	Surviving
	RamgarhPanjoopur	0.44	0.31	0.99	0.17	Surviving
	Dhanipur	0.41	0.49	0.82	0.20	Surviving
	KasbaKol	0.31	0.28	0.63	0.09	Adapting
	Alinagar	0.20	0.62	0.58	0.23	Surviving
	Sarsaul	0.34	0.41	0.72	0.16	Surviving
	HaivatpurFagoi	0.27	0.26	0.63	0.09	Adapting
Intermediate urban fringe	Barautha	0.32	0.28	0.40	0.02	Adapting
	Jarauth	0.40	0.58	0.30	−0.06	Adapting
	Kothia	0.28	0.71	0.19	−0.07	Adapting
	Nagaula	0.45	0.46	0.25	−0.09	Accumulating
	BhankriKhas	0.48	0.63	0.31	−0.11	Accumulating
	Digsi	0.60	0.39	0.23	−0.14	Accumulating
	Imlani	0.32	0.60	0.11	−0.13	Accumulating
	Karsua	0.40	0.40	0.32	−0.03	Adapting
	KishanGarhi	0.39	0.71	0.15	−0.17	Accumulating
	Jatpura	0.34	0.37	0.19	−0.06	Adapting
	Alipur	0.41	0.67	0.20	−0.14	Accumulating
	RafipurSiya	0.32	0.52	0.24	−0.04	Adapting
	Badwamni	0.33	0.64	0.26	−0.05	Adapting
	Bhartua	0.30	0.68	0.18	−0.08	Accumulating
	MadanpurChhabila	0.29	0.65	0.16	−0.08	Accumulating

(continued)

Table 7.2 (continued)

Zone	Name	Asset index	Sensitivity index	Exposure index	Sustainable livelihood index	Stages of livelihood ladder
Distant urban fringe	Maimari	0.42	0.44	0.00	−0.18	Accumulating
	Arrana	0.41	0.35	0.17	−0.09	Accumulating
	Chuharpur	0.38	0.49	0.34	−0.02	Adapting
	Ahamadpuri	0.47	0.29	0.26	−0.06	Adapting
	Tejpur	0.32	0.32	0.14	−0.06	Adapting
	SisroiNaglaDhak	0.33	0.35	0.14	−0.07	Adapting
	Peepali	0.31	0.58	0.39	0.04	Adapting
	DaudGarhi	0.34	0.53	0.21	−0.07	Adapting
	Navalpur	0.27	0.46	0.27	0.00	Adapting
	GadaKhera	0.38	0.25	0.06	−0.08	Accumulating
	Sikargarhi	0.38	0.33	0.16	−0.07	Adapting
	Jagatpur	0.33	0.44	0.25	−0.04	Adapting
	Pratappur	0.28	0.42	0.17	−0.05	Adapting
	JangalGarhi	0.30	0.24	0.18	−0.03	Adapting
	Padki	0.32	0.38	0.10	−0.08	Accumulating
	Daupur	0.27	0.35	0.09	−0.06	Adapting
	Bhataula	0.26	0.42	0.20	−0.03	Adapting
	Parauri	0.52	0.29	0.15	−0.11	Accumulating

Source Field Survey, 2013 and Census of India, 2011

are continuously under monetary stress. Many households have adopted crop diversification in which they adopted vegetable cultivation which has substantial demand in the city market, but in this strategy there is risk of crop failure because vegetables are susceptible to deceases and pests. Though these households have the advantage of city proximity but they also have to bear the urban impacts. Since the study area is more agrarian based and still under process of transforming, thus they are unable to cope with the stress of urbanization.

The evaluation of livelihood status in Aligarh urban fringe, the study finds that the better status on livelihood ladder (accumulating stage) is found in only 13 of the sampled 44 villages. This accumulating stage has the ability of absorbing livelihood shocks. But interestingly none of the immediate urban fringe village was found in this stage, while 8 from intermediate and 5 from distant urban fringe came in this stage (Table 7.2). This suggests the immediate urban fringe is vulnerable to shocks because they have to face the brunt of urban forces.

On the second stage of livelihood ladder (adapting stage), livelihood situation is improving but still vulnerable to extreme shocks (Table 7.3). Out of the 44 sampled villages, 25 came under this stage, interestingly higher numbers of villages from distant urban fringe were found to be in adapting stage. It suggests that greater distance from city restrict them from advantages of urban resources but more significantly the intensity of stress exerted by urban forces is also reduced. The households adapt the strategy of male selective migration to take advantage of urban resources and distance from city give the cushion from adversities of urban forces.

Table 7.3 Stages of Aligarh's urban fringe villages in livelihood ladder

Accumulating	Life is going well Can cope with most external stress/urbanization Range of choices, very flexible Works in formal economy Accumulating assets	Immediate urban fringe—0 villages (0 %) Intermediate urban fringe—8 villages (53.3 %) Distant urban fringe—5 villages (27.7 %)
Adapting	Life isn't bad Vulnerable to shocks (reduced benefits access) More choice and flexibility but still limited Has moved into formal economy (often low paid)	Immediate urban fringe—5 villages (45.5 %) Intermediate urban fringe—7 villages (46.6 %) Distant urban fringe—13 villages (72.2 %)
Surviving	Life is a constant battle—only just surviving Very vulnerable to any external shocks No choice or flexibility Unable to cope with shocks because of lower assets	Immediate urban fringe—6 villages (54.5 %) Intermediate urban fringe—0 villages (0 %) Distant urban fringe—0 villages (0 %)

Source Field Survey, 2013 and Census of India 2011

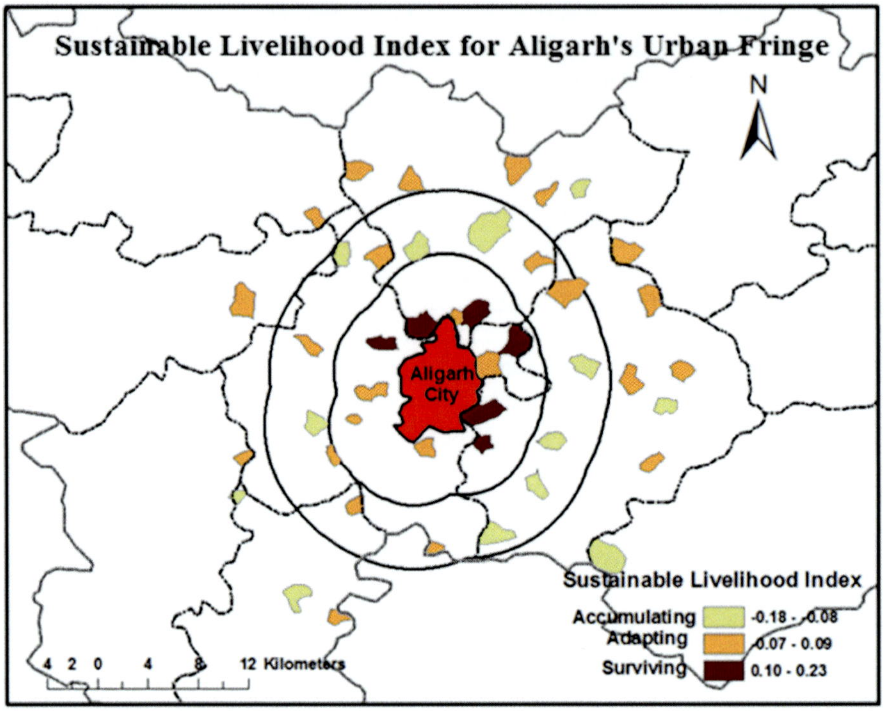

Fig. 7.6 *Source* Field Survey, 2013 and Census of India, 2011

The third step of livelihood ladder (surviving stage) is characterized by higher degree of vulnerability and households found to be least prepared to cope with livelihood shocks. The study finds that immediate urban fringe villages were the main sufferer(Fig. 7.6). The households in immediate urban fringe are found to be vulnerable to urban stress because of their poor economic and social capital. They were unable to take advantage of city proximity and their lesser assets restrict them of livelihood choices and flexibility.

References

Abbas K et al (2013) Household responses to livelihood transformation in peri-urban Kumasi. J Sustain Dev 6(6):121–136

Aberra E, King R (2005) Additional knowledge of livelihoods in the Kumasi peri-urban interface (KPUI), Ashanti Region, Ghana. Development Planning Unit, and University College London. Retrieved October 2, 2010, from http://www.nrsp.org.uk/database/documents/2813.pdf

Allen A, da Silva NLA, Corubolo E (1999) Environmental problems and opportunities of the peri-urban interface and their impact upon the poor. Development Planning Unit, University College London

Banu N, Fazal S (2013) Development of infrastructural facilities in public sector on the PUI of Aligarh City: a regional perspective from North India. J Infrastruct Dev 5(2):151–168
Banu N, Fazal S (2014) Agricultural transformations in the peri-urban interface of a developing economy: a case study. Man Dev 36(2):31–47 (Centre for Research in Rural and Industrial Development (CRRID))
Brush JE (1968) Spatial patterns of population in Indian cities. Geogr Rev 58:362–691
Chambers R, Conway G (1992) Sustainable rural livelihoods: practical concepts for the 21st century. Institute of Development Studies, UK
Department for International Development (DFID) (1999) Sustainable livelihoods guidance sheets
Food and Agriculture Organization (2003) Overview of land value conditions. ftp://ftp.fao.org/agl/agll/docs/misc35e.pdf
Fazal S (2013) Land use dynamics in a developing economy. regional perspective from India. Springer Publication, The Neatherlands
Fazal S (2014) Peri urban livelihoods opportunities and challenges. Concept Pulication, New Delhi
Fischer G, Sun L (2001) Model based analysis of future land-use development in China. Agric Ecosyst Environ 86(3):163–176
Hahn Micah B et al (2009) The livelihood vulnerability index: a pragmatic approach to assessing risks from climate variability and change—a case study in mozambique. Glob Environ Change 19:74–88
IPCC (2001) Climate change 2001: impacts, adaptation, and vulnerability, contribution of working group II to the third assessment report. Cambridge University Press, Cambridge
Lewis GJ, Maund DJ (1976) The urbanization of the countryside: a framework for analysis. Geogr Ann Ser B Hum Geogr 58(1):17–27
Lyon D (1983) Development of the urban-rural fringe: a literature review. Research and working papers No. 3. Institute of Urban Studies, University of Winnipeg, Winnipeg
May C et al (2009) The sustainable livelihoods handbook: an asset based approach to poverty. Published by Church Action on Poverty and Oxfam GB
Owusu G, Agyei J (2007) Changes in land access, rights and livelihoods in peri-urban Ghana: the case of Accra, Kumasi and Tamale metropolis. ISSER, Accra
Sharma D (2010) Who will feed Uttar Pradesh? In other words, who will feed India in the days to come? Ground reality: understanding the politics of food, agriculture and hunger. Retrieved from http://devinder-sharma.blogspot.in/
Shaw A (1999) Emerging pattern of urbanisation in India. Econ Polit Wkly 34(16):17–30
Shaw A (2005) Peri-urban interface of Indian cities: growth, governance and local initiatives. Econ Polit Wkly 40(2):129–136

Conclusions

Aligarh is a relatively small city, compared to other cities of which literature is available but it is also a fast growing city in India, the issues examined here are certainly not limited to Aligarh city and many of the features can be traced with other cities as well. It is surrounded by a populous rural area with productive and rich agricultural hinterland. Such conditions give rise to many conflicts between rural and urban economies, values and people. They can also generate mutually beneficial complementarities in the rural and urban spheres. The net outcome is highly selective for different groups and different locations. The present study was spread over 44 villages, where a total of 757 households were sampled for in-depth analysis.

The literature on rural–urban fringe is in large volume but because of its dynamic nature the definition and delimitation is rather inconsistent. This is largely due to differences in site and situation which results in different kind of urbanization process and the resultant impacts on urban fringe. The study considers urban fringe as transitory zone, where combination of rural and urban activities meet. The urban fringe is region of change and not a distinct boundary, where the process of change is crucial.

The development in city and the expansion of city strongly influences the peripheral rural settlements in terms of natural resources and people, both physically and in a socio-economic sense. Its influence stretches far beyond the immediately adjacent area. The implications of it were felt on the people who have lived there for generations as well on that of the migrants who came here late. This study was for assessing livelihood of people living in city surrounding (the urban fringe). Livelihood is a method to acquire food, income, property or inherit resources; it includes all the activities forming the means of generating a living.

A fundamental characteristic of urban fringe livelihood is their ability to adapt in order to survive. The feature of urban fringe livelihood includes change over time and adaptation to evolving circumstances and this is an ongoing process which does not remain static but varies from one season or from one year to the next. The assets can be built up, eroded or instantaneously destroyed as available activities and access to resource and opportunities may fluctuate seasonally and across the year.

The study finds that there was clear signs of modernization and transformation of economic activities in the Aligarh's urban fringe. The traditionally dominant agricultural activities are under increasing pressure and new possibilities are

N. Banu and S. Fazal, *Livelihood and Wellbeing in the Urban Fringe*,
The Urban Book Series, DOI 10.1007/978-3-319-39660-6

emerging for market-oriented agriculture. There were great differences in access to resources and opportunities leading to new patterns of livelihood. Significantly, options for non-agricultural occupations among the urban fringe households are enhanced because of their proximity to the expanding city.

The pull factor of urban infrastructure and greater employment opportunities attract migrants from the surrounding villages. There is phenomenal increase in linkages and the village economy gets tied up with the growing city. With time and improvements in the communication and transport facilities, the daily movements from the village to the city in order to supply the different commodities have increased considerably. The rural settlements in the urban fringe are no longer inhabited by traditional rural population only, but they also inhabit people who spend most part of the day in city.

The study has found that the livelihood in Aligarh's urban fringe can be categorized into two broad categories—noncash-based livelihood and cash-based livelihood. The noncash-based livelihood, where households are engaged with food production and fuel or fodder collection from common land. But this form of livelihood has been almost wiped out in Aligarh's urban fringe. Beside the dependency on common land, another type of noncash-based livelihood is manual scavenging of village roads. In the urban fringe of Aligarh city, especially women from poor families are forced into this job. They receive miserable food and clothes in return. The study finds that in Aligarh's urban fringe, the urban elements linked with monetized economy had affected rural elements linked to natural resources resulting in their gradual depletion. Moreover, poor people who earlier depended on these natural resources from common land have lost control over it. These changes in urban fringe have adversely affected their noncash-based livelihood in particular.

The cash-based urban fringe livelihoods are categorized in to four broad categories. These are livelihoods based on casual employment, livelihoods based on business and trade, livelihoods based on farming activity and livelihoods based on dairy farming. Cash-based casual livelihood in Aligarh's urban fringe has many dimensions and most of the lesser educated and lesser skilled workers belong to this category. There is increasing demand for labour-oriented jobs in urban fringe. The employment opportunity of this kind is typically casual, short term and seasonal in nature fetching low and varying wages.

This kind of labour work can be divided into two categories: labouring in agricultural sector, where the wages are comparatively lesser, interestingly, there is considerable shortage of labour in the agricultural sector but its remuneration does not increase to the levels of labour in non-agricultural sector of urban areas. This kind of urban fringe livelihood is characterized by its highly seasonal nature and this seasonal migration to them is like "*Paid holiday*". Importantly, this sector is not gender biased, in fact women are found to have almost equal opportunity for work, but the wages are lesser than their male counterparts.

The livelihood of labouring outside agricultural sector is a cash-based livelihood opportunity in unorganized urban sector. The urban fringe offer opportunity to sell labour for remittance in cash. This livelihood option gave an opening for rural labourers to move away from primary sector and earn relatively higher wages.

However, this option is typically casual, seasonal, short term and 'by day' employment. The urban proximity is an important requirement for this kind of livelihood. Moreover, this employment is highly gender biased with male dominance.

Another livelihood option in Aligarh's urban fringe is cash-based off-farm livelihood in tertiary sector which requires some skills (most of them are craft based, artisanal activities or providing lower level services), small time trading is also one of the livelihood options for this section. This option is directly linked to urban economy and thus, vibrant with proximity to the city. This option is popular with younger people with some level of education and who have distanced themselves from labour-oriented agricultural activities. They prefer to work and stay in urban area and take advantages of urban infrastructure and amenities.

The livelihood as crop cultivation is the most common and widely practiced option in the Aligarh's urban fringe. This option in the urban fringe is found to have fair degree of commercialization and households were found to be producing farm products for the market to get cash returns. Significantly, distance from city has crucial bearing on nature of farming activity. The sample survey suggests that closer to the city have higher degree of commercialization and it decreases as we move away from city.

The farmers also found exploring newer options for drawing maximum returns. Especially, producers from *immediate* urban fringe acting in urban system were found to be exploring most for possibilities of better marketing and management for their products. The urban fringe producers also showed interests in extension services offered by governmental agencies for upgrading their traditional skills. This modification was helping them in judiciously utilizing their stressed resources in urban fringe.

The study area lies in an agriculturally prosperous region of the country, where inputs of green revolution were adopted successfully. That is why the farms of Aligarh's urban fringe have reported higher level of application for agricultural inputs. The basic inputs of green revolution such as irrigation, high yielding varieties of seeds, chemical fertilizers etc., were widely used by the sampled cultivators. The cropping pattern was found to be greatly influenced by urban proximity and some crops such as vegetables, peas, *arhar,* flowers, fruits and *chara* etc., have substantial acreage, these commodities are basically grown to meet the urban demands. Whereas, farms at *distant* urban fringe have shown subsistence orientation in their cropping pattern, as the cropping pattern was dominated by food grains.

Livestock, particularly dairying, emerges as the most important natural resource-based livelihood option in the urban fringe. The study assumes that the livestock-based livelihoods are greatly influenced by urban growth. This livelihood option has inherent urban fringe characteristics as households afford multiple options where the men can work in the farm as well as off-farm (in some urban activity or as middlemen), similarly, women can perform the men's as well as their own tasks of raising animals. Buffalo is the preferred milking animal, as consumers prefer its milk because of its high fat content. The distance from urban area is critical for dairy activity because milk being a perishable commodity and with no refrigeration or preservation techniques are used, this activity thrives closer to the city. Other livestock (except goats for some people) do not form a substantial part of livelihood in the urban fringe. Goats, though still small in number, seem to be increasing. But it is a more general than exclusive urban fringe phenomena.

The study for assessing the status of land resource in Aligarh's urban fringe finds severe *urban shadow effect* on it. There is substantial conversion of crop land to urban expansion and also for tree plantation and vacant lands in the urban fringe. Farmers in the urban fringe subject to present or short-term future development pressure, begin to devote less capital for farm maintenance and management resulting in lower productivity from the land, which ultimately results in selling these lands to property dealers. The net effect of these forces cause a much greater loss of cropland to urban uses than the figures normally given for land transferred to urban uses indicate.

The sample survey further found landowners disposing their lands in response to the urban forces differently because of diverse economic and social background. Large land owners were found to be more responsive towards the urban forces. The reason was that they have distanced themselves from direct agricultural activity. This poses managerial and operational problems in farming activity. Medium land owners were found to be resisting to urban forces and still associate themselves with farming community, but they are also the progressive and dynamic section of the society, where they change occupation and this instigate them to dispose off their agricultural lands. Small landowners are also responsive to urban forces for disposing off their agricultural land. But their decision is more due to their poor and miserable economic background.

The study also investigated the impact of city on urban fringe population; this impact was found to be quite noteworthy on their economic, social, infrastructural and natural environment. There were strong evidences of the transformation in economic environment in the urban fringe. This zone is characterized by the coexistence of urban (mainly secondary and tertiary activities) and rural (mainly primary or natural resource based) activities. The study finds that urban fringe is a place with increased opportunities as the urban-based economic activities consisting of secondary and rapidly expanding tertiary sector. The urban fringe also benefited from the larger urban market. The city with increasing population creates increased demand mainly for primary (nature based) products. The easy access (personal contact with consumers) and monetized urban economy is of great advantage to these urban fringe producers. This urban monetized economy, larger market, greater opportunities and higher-assured wages have led urban fringe households to diversify their livelihood for better and improved living. There have been evidences of occupational transformations and this phenomenon was more prominent among younger generation and clear urban fringe character particularly strong with proximity to the city where greater choices and opportunities exist.

The study assessed the household quality of life of the sampled villages in the urban fringe of Aligarh city; a Household Quality of Life Index (HQLI) has been made. The sample survey finds that there was a marked influence of city life on the life style and quality of life of urban fringe households. There was significant improvements in the urban fringe households which have become more urbanized and the quality of life have also improved considerably. Most of this improvement in urban fringe is because of inflow of money through sale of land or income from secondary and tertiary sector. This improvement is visible from their houses (which are mostly pucca, constructed using bricks and cement). Distance from market, both physical and cognitive, significantly influence individuals, economic prospects. By alternatively facilitating or limiting access to market based opportunities, distance can inverse critically between effort and reward.

There have been improvements also in sanitation, ventilation, electricity connections, use of fuel, availability of modern gadgets etc. However, all these improvements decreases with distance, as immediate urban fringe have better housing condition where as distant urban fringe has relatively poor condition. The effects of the differences in earning potentials become visible at asset ownership of the household. The scores in asset possession index showed the variations in each type of assets examined was owned by a greater proportion of households in the immediate urban fringe compared to as the distanced zones reflecting the effects of distances in earning opportunities. There was substantial differences in asset holdings were also visible within the urban fringe areas. Every asset type examined was possessed by larger share of households in immediate urban fringe as compared to distant urban fringe villages.

The effects of the differences in economic improvement become visible in the Housing Condition Index (HCI), Asset Possession Index (API) and in Household Quality of Life Index (HQLI). There have been improvements also in housing condition as well as in modern asset possession. However, this improvements increase with city accessibility, as immediate urban fringe have better housing condition and higher asset possession where as distant urban fringe has relatively poor housing condition and lower asset possession. In household quality of life index, 7 villages (64 %) out of 11 villages in the immediate urban fringe reported of having high quality of life and rest 4 villages (36 %) having medium quality of life. But as we move out to distant urban fringe 15 villages (83 %) out of 18 villages falls under low quality of life and rest 3 villages (17 %) having medium quality of life but not a single one reported in the category of high quality of life. This points towards impact of city proximity in the improvement on quality of life.

The study also found that city impact have significantly transformed social environment in the urban fringe. The urban fringe is characterized by the intermixing of urban way of life and rural way of living. But this intermixing is in some way is constrained by cultural differences. Urban Fringe is a zone of high growth in its population marked with intensification of concentration of population. There are also signs of improving status of women who were actively participating in economic activities, managing the household as well as taking care of household resources too. The education, political empowerment and economically activeness have made them secure. Another important change which the urban fringe has witnessed is awareness and popularization of education among the masses. Education has started to transform the urban fringe society. Even the poor and uneducated households have also started to push their children for pursuing education. Urban fringe has also witnessed transformation in family structure. The families with changes in occupational structure are increasingly becoming nuclear. The study also finds a marked influence of city life on the life style and living standard of urban fringe households. The urban fringe households have become more urbanized. The inflow of money through sale of land and incomes from secondary and tertiary sector has contributed for this improvement, but these improvements decreases with distance.

The study also finds that urban fringe is a place of new social class which has increasing elements of urban characteristics and decreasing rural elements. The society has better education levels, smaller family size but marked gender biasness. Although they showed gender biasness among kids but general status of women has improved. The households have showed association with diverse economic activity. This multi-activity household is survival response to urban monetized economy to generate higher incomes required for sustenance. This urban monetized economy has also blurred the traditional caste-based social system. The urban fringe has social class, where social status is determined by household income. This income-based social evolvement has also transformed the urban fringe society in to materialistic society which has predominant urban values and way of life. The urban fringe residents do not have much affinity with one another and the sense of social responsibility among the people has been reduced.

The study also finds urban fringe a place with contradicting perception as on one hand there have been positive improvements as stated above but respondents were also concern as their economic, social and cultural activities have been greatly transformed, which in their view have changed their life style and eroded their social and cultural values. However, this dichotomy was not evident among younger generation. Interestingly, the distant urban fringe respondents on the other hand complain of difficulties in accessing urban benefits, they want rapid urban development and expansions so that they come closer to the city and get benefits of urban amenities and facilities.

The city impact on urban fringe also helped in improvements of infrastructural facilities. Urban impact on urban fringe is both in terms of ‘site’ and ‘situation’. Here, ‘site’ is referred to the physical and cultural characteristics and attributes of the city. While, ‘situation’ specifies the external relations of the locale. In the context of a urban fringe, a change in situation usually precedes a change in the site. This is evident from the improvements in the infrastructural facilities, amenities and their access in the urban fringe. Better access to a location strengthens the external relations of that place leading to changes in its site. The transformation in availability and access to infrastructural facilities and amenities in the urban fringe has greatly improved. Moreover, there is also an improvement in general awareness for accessing and using these available facilities.

The study also assessed the status of infrastructural facilities for sampled villages in the urban fringe of Aligarh city, an Infrastructural Facility Index (IFI) was made up of four distinctive indices such as transport facility index (TFI), education facility index (EFI), health facility (HFI) and marketing facility index (MFI).

The analysis to assess infrastructural facilities in the villages of Aligarh urban fringe showed that higher scores were found in the villages of immediate fringe and with distance lower scores were found. The analysis shows that only two villages reported very high infrastructural facility scores and both of them were from immediate urban fringe. Another six villages were placed in high category of which three each were from immediate and intermediate urban fringe zone. On the other hand, very low and low scores for infrastructural facility index were reported higher in distant villages as out of eighteen sampled villages, eight villages categorized in very low and four villages in low category. This clearly indicated that distance from city make remarkable effects on existence of infrastructural facilities in urban fringe.

Cities are dynamic human artifact, which constantly undergo structural change, redevelopment and growth. Such processes also involve changes in urban relationships with the surrounding territory, most conspicuously on their outskirts. The urban area offers greater opportunities in its surrounding area but it also exert negative shadow effects. The increasing pressures or stress from urban expansion pose a serious threat to urban fringe livelihood because this process affects fertile crop land (the dominant natural capital) on which most households in the urban fringe area depend. In the study area, due to city's expansion, there was significant loss of fertile crop land over the past four decades. Since the study area is situated on the agricultural belt of India due to ferrite alluvial soil of Ganga–Yamuna Doab and most of the people engaged in agricultural activities. Thus any loss to agricultural land has its bearing on livelihood of people.

The analysis of composite asset possession, household from intermediate urban fringe have reported highest scores among the sampled households followed by distant and immediate urban fringe. The intermediate urban fringe has greater asset possession among human, social, natural and financial capital. It suggests that intermediate urban fringe households are taking advantage on both counts. On one hand they are better placed as the urban resources are in their reach while immediate urban fringe take the maximum impact of urban shadow effects, protecting from the ill effects (shock) of urban forces.

The status of immediate urban fringe households in term of composite asset possession is poor although, household from immediate urban fringe have reported higher scores for physical capitals (consists of amenities and infrastructural facilities) because of urban proximity. This lower scores for Composite Asset Index is mainly because lower social, natural and financial capital.

The status of asset possession among distant urban fringe is also satisfactory as they have higher scores on human and financial capital and moderate scores for natural, social and physical capital. The household in distant urban fringe are disadvantageous because of their farness from city but in a predominantly agriculture based livelihood society, it does not impact much and on the contrary this distance protect them from urban shadow effects and shocks.

Finally, the study finds that urbanization process is quite prominent in India and it has significant influence on urban fringe. The study area represents a predominantly traditional agrarian society, where agriculture is the 'culture'. Here any change has to pass through agriculture not only in economic sense but also in social and cultural sense.

Land is the most important asset in urban fringe, irrespective of its productivity, its mere possession ensure security to their livelihood. The importance of land resource is so much so that the urban fringe households who have disposed off their land in return to handsome monetary benefits are found to be struggling to secure their livelihood.

The immediate urban fringe households where in natural asset (land) is diminishing under the urban forces creating severe stress on livelihood. To cope these, households look for livelihood in urban sector, but due to their limited educational and technical attainments, they barely manage themselves to surviving stage of livelihood ladder. The landowners in this zone have gained assets through high land rent and diversification of their cropping pattern to meet urban demands, it is their coping strategy. Overall immediate urban fringe households were found on lower step of livelihood ladder despite the city proximity. This suggests that rural households in present form are unable to take advantage of urban monetized economy.

The intermediate urban fringe households were found better placed on livelihood ladder in the study region. They take advantage of retaining their natural capital (land resource), diversifying their cropping pattern, diversifying their livelihood and adopting male specific migration to enhance assets. Their intermediate site within urban fringe is advantageous in two ways; first most of the urban ill effects are absorbed by immediate urban fringe while they take advantage of city proximity also.

The distant urban fringe is still continuing with more or less rural dominated livelihood, they have limited access to urban economy but also away from urban shadow effects. They possess natural capital but its value is lesser, similarly degree of diversification in livelihood and cropping pattern is also lower resulting lesser asset possession.

Suggestions:

- India is urbanizing, so is the expansion of urban area is taking place. All this urban expansion is occurring on peripheral areas around the existing settlements—urban fringe (UF). There is an urgent need to have appropriate planning and adequate and specific policies for developmental issues for this urban fringe.
- The urban fringe is a very important zone which provides not only primary products but now also secondary commodities and tertiary services to the city. but the marketing system and channels are poor and grossly inadequate. Moreover, the marketing system is urban biased which favours the middlemen (*dalal*) who gets major share of profits from the transactions. UF producers are victims of this faulty system and suffer from poor payments and are also have to bear the burden of taxes and commissions. There is an urgent need to develop a balanced marketing system to safeguard the interests of urban fringe producers.
- The youth of urban fringe requires special attention, as this young human resource is in dilemma. They have distanced themselves from traditional

labour-oriented primary and secondary activities and their abilities do not absorb them in desired urban-based tertiary sector. This way more and more young people remain idle and unemployed. They could be trained and utilized in marketing of urban fringe commodities.

- One of the intractable environmental issues is the inevitable conversion of agricultural land to urban uses and land degradation. The availability of larger tracts of relatively undeveloped and cheaper land in the city's urban fringe makes it environmentally susceptible. In other words, negative environmental externalities are imposed on the urban fringe. This needs to be checked immediately as in the name of development humanity can not afford to loose indiscriminate fertile agricultural land. Because it is not only about the revenue loss due to loss of agricultural land but it is also an issue of livelihood of the people who were associated with that land.
- Livestock development should become the core of livelihood programmes in the urban fringe context. The study finds that land is lost for non-agricultural purposes, eroding the natural resource base of the community. Livestock emerges to be more important a livelihood than agriculture as one approaches the city. This means, more livestock have to be managed with less land. But this inverse proportion of livestock and land can not go beyond a limit without both being affected. Fodder resources need urgent attention in preserving or promoting cropping patterns that produce sufficient fodder. The vacant land can also be utilized for producing fodder.
- Livestock, particularly dairying, emerges as the most important natural resource-based livelihood option for the urban fringe households. But healthcare of animals is an area of concern despite better economics and accessibility to services. The dairy producers need to appraise about the benefits and importance of these facilities. The community can be empowered to take care of these services themselves. Training of local youth as para-vet to perform vaccinations, artificial insemination, pregnancy diagnoses and first aid will be of immense benefit.
- The effect of urbanization is real as far as dairy farming is concerned. There are distinct dissimilarities between the immediate and distant urban fringe villages in composition of livestock population, livestock management, milk marketing and consumption. Development programmes need to recognize that the urban fringe situation differs from rural and urban ones.
- Agriculture is still an important and dominating livelihood option, but in urban fringe it suffers adversely due to urban forces. Agricultural activity requires special attention and protection so that it does not succumb to urban forces.
- There is a critical issue which requires immediate attention. This is about growing menace of application and adulteration of harmful chemicals into milk, vegetables and fruits. This needs to address as most of the rural producers are ignorant of its ill effects on human life. The producers should be guided and made aware of their detrimental consequences and desist them from these illegal acts.

Appendix A
Aligarh Urban Fringe: Household's Income Generated from Different Sectors

Chi-square tests			
	Value	df	Asymp. Sig. (2-sided)
Pearson chi-square	17.707[a]	2	0.000
Likelihood ratio	17.337	2	0.000
Linear-by-linear association	17.368	1	0.000
N of valid cases	757		

[a]0 cells (0 %) have expected count less than 5. The minimum expected count is 54.38

Symmetric measures			
		Value	Approx. Sig.
Nominal by nominal	Phi	0.153	0.000
	Cramer's V	0.153	0.000
N of valid cases		757	

N. Banu and S. Fazal, *Livelihood and Wellbeing in the Urban Fringe*,
The Urban Book Series, DOI 10.1007/978-3-319-39660-6

Appendix B
Aligarh Urban Fringe: Sector Wise Working Age Group

Chi-square tests

	Value	df	Asymp. Sig. (2-sided)
Pearson chi-square	84.205[a]	2	0.000
Likelihood ratio	85.190	2	0.000
Linear-by-linear association	53.287	1	0.000
N of valid cases	1703		

[a]0 cells (0 %) have expected count less than 5. The minimum expected count is 63.09

Symmetric measures

		Value	Approx. Sig.
Nominal by nominal	Phi	0.222	0.000
	Cramer's V	0.222	0.000
N of valid cases		1703	

N. Banu and S. Fazal, *Livelihood and Wellbeing in the Urban Fringe*,
The Urban Book Series, DOI 10.1007/978-3-319-39660-6

Appendix C
Aligarh Urban Fringe: Disposal of Produced Crop from Sampled Households

Chi-square tests

	Value	df	Asymp. Sig. (2-sided)
Pearson chi-square	186.485[a]	8	0.000
Likelihood ratio	217.319	8	0.000
Linear-by-linear association	21.648	1	0.000
N of valid cases	457		

[a]3 cells (20.0 %) have expected count less than 5. The minimum expected count is 2.98

Symmetric measures

		Value	Approx. Sig.
Nominal by nominal	Phi	0.639	0.000
	Cramer's V	0.452	0.000
N of valid cases		457	

N. Banu and S. Fazal, *Livelihood and Wellbeing in the Urban Fringe*,
The Urban Book Series, DOI 10.1007/978-3-319-39660-6

Appendix D
Transformation of Operational Land Holdings Under Different Crops

Crops	Land (in ha) 2000–2001	Land (in ha) 2000–2006	Increase/decrease	Growth rate (%)
Cereals	104729	114445	+9716	+9.27
Vegetables	2830	2988	+158	+5.58
Flower	5	299	+294	+58
Fruits	2152	1548	−604	−28.06
Fodder	10456	582	−9874	−94.43
Total operational land	82616	81323	−1293	−1.56

Source Agricultural Census of India

N. Banu and S. Fazal, *Livelihood and Wellbeing in the Urban Fringe*,
The Urban Book Series, DOI 10.1007/978-3-319-39660-6

Appendix E
Aligarh Urban Fringe: Dairying Activity is Thriving in Immediate Urban Fringe

Chi-square tests

	Value	df	Asymp. Sig. (2-sided)
Pearson chi-square	18.801[a]	2	0.000
Likelihood ratio	18.976	2	0.000
Linear-by-linear association	16.426	1	0.000
N of valid cases	757		

[a]0 cells (0 %) have expected count less than 5. The minimum expected count is 58.55

Symmetric measures

		Value	Approx. Sig.
Nominal by nominal	Phi	0.158	0.000
	Cramer's V	0.158	0.000
N of valid cases		757	

N. Banu and S. Fazal, *Livelihood and Wellbeing in the Urban Fringe*,
The Urban Book Series, DOI 10.1007/978-3-319-39660-6

Appendix F
Aligarh Urban Fringe: Dairying Activity is a Joint Family Venture

Chi-square tests

	Value	df	Asymp. Sig. (2-sided)
Pearson chi-square	82.603[a]	2	0.000
Likelihood ratio	80.257	2	0.000
Linear-by-linear association	67.350	1	0.000
N of valid cases	757		

[a]0 cells (0 %) have expected count less than 5. The minimum expected count is 46.91

Symmetric measures

		Value	Approx. Sig.
Nominal by nominal	Phi	0.330	0.000
	Cramer's V	0.330	0.000
N of valid cases		757	

N. Banu and S. Fazal, *Livelihood and Wellbeing in the Urban Fringe*,
The Urban Book Series, DOI 10.1007/978-3-319-39660-6

Appendix G
Aligarh Urban Fringe: Milk Sales and Milk Market

Chi-square tests			
	Value	df	Asymp. Sig. (2-sided)
Pearson chi-square	140.986[a]	8	0.000
Likelihood ratio	154.610	8	0.000
Linear-by-linear association	3.223	1	0.073
N of valid cases	490		

[a]0 cells (0 %) have expected count less than 5. The minimum expected count is 5.23

Symmetric measures			
		Value	Approx. Sig.
Nominal by nominal	Phi	0.536	0.000
	Cramer's V	0.379	0.000
N of valid cases		490	

N. Banu and S. Fazal, *Livelihood and Wellbeing in the Urban Fringe*,
The Urban Book Series, DOI 10.1007/978-3-319-39660-6

Appendix H
Aligarh Urban Fringe: Mean Retail Price of Produced Milk

Descriptive

Milk rate

	N	Mean	Std. deviation	Std. error	95 % confidence interval for mean		Minimum	Maximum
					Lower bound	Upper bound		
immediate urban fringe	226	25.8319	14.82724	0.98629	23.8883	27.7754	0.00	45.00
intermediate urban fringe	142	18.9366	13.45558	1.12917	16.7043	21.1689	0.00	38.00
distant urban fringe	122	16.5410	11.20002	1.01400	14.5335	18.5485	0.00	30.00
Total	490	21.5204	14.18158	0.64066	20.2616	22.7792	0.00	45.00

Test of homogeneity of variances

Milk rate

Levene statistic	df1	df2	Sig.
6.439	2	487	0.102

N. Banu and S. Fazal, *Livelihood and Wellbeing in the Urban Fringe*,
The Urban Book Series, DOI 10.1007/978-3-319-39660-6

Appendix I
Aligarh Urban Fringe—Mean Milk Productivity

Milk production

Descriptives

Total milk productivity per day

	N	Mean	Std. deviation	Std. error	95 % confidence interval for mean		Minimum	Maximum
					Lower bound	Upper bound		
Immediate UF	226	11.8673	8.23975	.54810	10.7872	12.9473	0.00	40.00
Intermediate UF	142	9.6338	7.53947	.63270	8.3830	10.8846	0.00	30.00
Distant UF	122	8.6557	6.96481	.63056	7.4074	9.9041	0.00	30.00
Total	490	10.4204	7.84583	.35444	9.7240	11.1168	0.00	40.00

Test of homogeneity of variances

Total milk productivity per day

Levene statistic	df1	df2	Sig.
1.187	2	487	0.306

Total milk productivity per day

Tukey HSD[a,,b]

Zone	N	Subset for alpha = 0.05	
		1	2
Distant UF	122	8.6557	
Intermediate UF	142	9.6338	
Immediate UF	226		11.8673
Sig.		0.512	1.000

Means for groups in homogeneous subsets are displayed
[a]Uses Harmonic Mean Sample Size = 152.565
[b]The group sizes are unequal. The harmonic mean of the group sizes is used. Type I error levels are not guaranteed

N. Banu and S. Fazal, *Livelihood and Wellbeing in the Urban Fringe*,
The Urban Book Series, DOI 10.1007/978-3-319-39660-6